Seismic Exploration Fundamentals

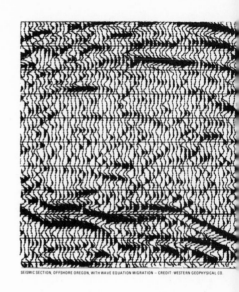

SEISMIC SECTION, OFFSHORE OREGON, WITH WAVE EQUATION MIGRATION – CREDIT: WESTERN GEOPHYSICAL CO.

Seismic

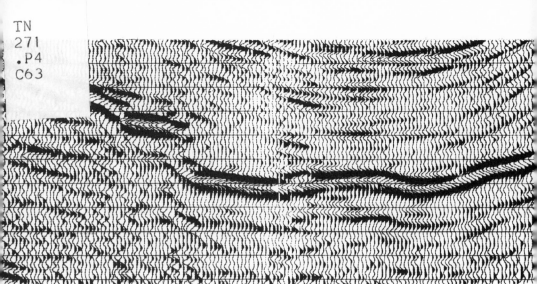

Exploration Fundamentals

The use of seismic techniques in finding oil

J. A. COFFEEN

PennWell Books

Division of
PennWell Publishing Company
1421 S. Sheridan
Tulsa, Oklahoma 74112

Library of Congress Catalog Card Number: 77–95312
International Standard Book Number: 0–87814–046–8

4   5 — 82   81

Contents

Preface   ix

1
What It's All About   1

Terms 2
Echo-Ranging 5
Reflection of Sound 5
Group 10

Spread 11
Common Depth Point 14
Refraction of Sound 19
Records and Sections 22

2
Phenomena   29

Weathering 29
Normal Moveout 30
Multiples 35
Ghosting 41

Bubbles 43
Noise 45
Wavelet 47
Velocity Variation 48

3
Field Operations   51

Program Planning 51
Field Geophysics 51
Field Crew 52
Permitting 54
Surveying 54
Energy Source 56
Shot Hole 57
Pattern Holes 60
Surface Explosives 61
Gas Gun 62
Weight Drop 62
Vibrator 63
LVL Crew 65
Cable 65
Geophone 66
Recording 67
Offshore 69

Vessel 69
Surveying 70
Preplots 71
Shoran 71
Satellite 72
Other Survey Systems 74
Hydrophone 75
Marine Cable 76
Air Gun 77
Marine Gas Gun 78
Other Sources 78
Recording 79
The Signal and its Signature 80
Telemetry 81
Shallow Water, Surf, Mud Flats 81
Engineering Surveys 82
Gravity and Magnetometer 85

4

Digital Recording

Analog and Digital 86
Binary Numbers 87
Exponential Notation 89

Binary Gain and Floating Point 90
Recording Format 91
Multiplex 92

86

5 -

Data Processing

Demultiplex 93
Near-Surface Corrections 94
Static Corrections 98
100% Section 100
Single Trace Gather 102
CDP Gather 102
Stacking 103
Multiple Attenuation 105
Filtering 106
Frequency Display 107
TVF 108

Muting 108
Velocity Analysis 109
Signal Theory 115
Correlation 115
Convolution 117
Deconvolution 122
Fourier and Frequency Domain 124
Flattening 125
Dip Rejection 126
Processing Sequence 127
Tape Storage 129

93

6

Record Sections

Display 134
Color 135

Vertical Exaggeration 135
Compressed Section 137

132

7

Data Handling

Velocity Determination 139
Velocity Survey 140
Synthetic Seismogram 142
Synthetic Sonic Log 146
Modeling 148
True Amplitude 151
Dip Migration 151

Diffraction Collapse 157
Wave Equation Migration 159
Comments on Migration 160
Ties in Migration 162
Three Dimensional Migration 164
3-D Shooting 168

139

8

Interpretation

Identification 175
Continuity 176

Loops 180
Formations and Reflections 181

175

Faults 184
Angle of Faulting 189
Mapping Faults 190
Mapping 191
Contouring—Geological
  and Geophysical 192
Computer Contouring 204
Time Interval Maps 206

Accuracy vs. Correctness 206
±500 or 500± 208
Bright Spot 208
Flat Spot 211
Stratigraphy from
  Seismic Sections 213
Misleading Features 214
Interpretation Sequence 215

9

Refraction Exploration

218

Refraction 218

Sonobuoy 222

10

Old Data

224

Availability of Old Data 224
Value of Old Data 225

Forms of Old Records 226
Combining Old and New 231

11

Types of Areas

233

Large Structures 233
Small Features 234
Salt Solution Features 235
Salt Domes 236

Reefs 239
Stratigraphic Traps 241
Lithologic Variation 242

12

General Considerations

244

Participation and Proprietary 244
Survey Problems 244
Seismic Reliability 246
Geological-Geophysical
  Coordination 248
Factors in Program Planning 250

Continuing Program 252
Development of Oil Fields 253
Drill the Top 254
Most Important Factor 256
How to Find Oil 256

Appendix

259

Other Sources of Information 259
Educational Requirements 261
What is a Geophysicist? 263

For Non-Oil People 263
How We Find Oil by
  Seismic Exploration 264

Preface

This is a book on using seismic techniques to *find oil*. It is *not* on mathematical and theoretical aspects of seismic exploration.

It is a manual designed for:

Managers, who need an overall knowledge of seismic exploration, to work with geophysics and geophysicists;

Geologists, whose work is closely related to geophysics;

People in specialized fields of geophysics—data processors, field personnel, etc., who would like an understanding of how their specialties fit into the exploration effort;

Finally, new employees with a theoretical background in geophysics, who need to learn how that background is used in finding oil.

The theoretical side of geophysics is vital. Seismic exploration began because theoretical principles, particularly Huygens' Principle, were developed. It is now progressing at an almost headlong rate, partly because of research being carried out by oil companies and seismic contractors.

This book is concerned with an entirely different problem—that of finding oil in specific situations. This calls for the exploration attitude, competitive, affected by deadlines, including many aspects of the situation simultaneously.

J.A.C.

Acknowledgments

Thanks are due to a number of people and organizations, the companies and individuals that contributed illustrations and examples, several friends who have read all or part of the text and commented on it, other friends I've hounded with questions, and the various people who have typed from my notes with the strange abbreviations, notations between the lines, and cut-and-paste rearrangements—Terrill Coffeen in Calgary, Lois Butler in Jakarta, Lilian Low in Singapore, Janet Wartick in Houston, and Mary Helen Coffeen in all those places.

Seismic Exploration Fundamentals

1

What It's All About

Seismic exploration for oil is searching underground for oil by use of sound. The word "seismic" refers to vibrations of the earth, which includes both earthquakes and the kind of sound that penetrates well into the earth. This sound is in the frequency range of about 10-100 cycles/sec. It is right at the low end of the sounds people can hear. Human hearing is in the range of about 20-20,000 cps, so whether the person can hear seismic sound probably depends on who the person is.

A sound in the proper range will go deep into the earth and return, bringing information back with it. The depth investigated is from about 1000 ft. to as much as 10 miles.

There are two basic types of seismic exploration, reflection and refraction. Reflection involves sound that is reflected, echoed, from a rock layer. Refraction uses sound that travels along a rock layer for some distance and returns to the surface. Refraction was the first type of seismic exploration commonly used, but reflection has now supplanted it in all but a few specialized uses. Reflection is the type of exploration that will be the subject of almost all this book. Reflection exploration is conducted by artificially producing a sound at or near the surface of the earth, and recording the echoes from underground.

Of course, it isn't as simple as that. The sound has to be produced in the right way, the recording has to be done carefully and correctly; there are many fine points in the form of corrections and ways to improve the recordings.

In conventional seismic exploration and its major variations, here is what happens.

Lines are drawn on a map to indicate where seismic investigations are to be conducted. The lines are surveyed, and points along the lines marked on the ground. At the places marked, holes are drilled, and explosives placed in the holes.

1

Small sound detectors are placed on the ground, in a line near one of the holes, and are connected by electrical cable to tape recording equipment. The explosive in the hole is detonated and the sounds from the explosion, including echoes from underground layers of rock, are recorded on tape.

This continues along the straight line, and the other planned lines are treated similarly (Fig. 1-1).

There are variations of this procedure. Instead of an explosive in a hole, a vibrator may be placed on the ground, or a gas exploded in a chamber on the ground, etc. Seismic work is also done at sea, on a boat. There, the sound is made by suddenly releasing compressed air, or by some other method, from devices trailed in the water. The sound detectors are built into a cable that is trailed from the boat.

The tapes with data recorded are sent to a computer, especially designed for handling seismic data. The data on the tape is re-arranged, modified, enhanced, on a succession of other tapes. A paper display of the data is made from the final tape. This is in the form of a seismic section, showing the upper several miles of the earth along the line being investigated. It is composed of many vertical lines of wiggles showing the sound vibration of the earth caused by the explosions.

The seismic sections are interpreted, using the echoes to determine depths to rock layers. When this has been done for a whole set of lines, the depths to a rock layer are plotted on a map and contoured to show the configuration of the layer. Potential oil or gas traps may be indicated by the contours. A well may be drilled on this basis. The well *may* find oil or gas.

Terms

Any group of people in a specialized endeavor, necessarily develops some expressions of its own. Helpful as these expressions are to the specialists though, they make it more difficult for outsiders to understand. So, throughout this book, some stress will be put on the terms used in geophysics. To start with, a few will be introduced now. These first terms refer to the description just given of a seismic operation.

The arrangement of locations for the seismic operation is a *seismic program*, and the laying out is *program planning*.

The shooting of a program is sometimes called a *seismic survey*, a somewhat confusing expression, because of its similarity to *surveying*, the fixing of locations and elevations in the field.

Fig. 1-1

J. A. Thomas

The group of people in the field is a *seismic crew*, or seismograph crew, also a doodlebug crew, and the individuals *doodlebuggers*.

The firing of explosives or other means of producing the sound is called *shooting*. By association, the whole operation in the field is also called shooting.

A hole in which the explosive is fired is a *shot hole*. Its geographical location is a *shot point*. By analogy and habit, the location of any other source of sound for seismic use is a shot point. Usually nowadays, the sounds are produced at such close spacing that "shot point" has changed from meaning every point shot, to only those used in mapping data. That is, when compressed air or a gas explosion, etc. is used, each use is called a *pop*, and it may be arbitrarily decided that, say, every eighth pop will be called a shot point. Or, more appropriately, it may be called a *data point*.

The line of shot points is a *continuous line*, seismic line, or just a line.

The sound itself is referred to as *seismic energy*, seismic waves, or just energy.

The sound detectors are called *geophones*, seismometers, detectors, or in field terminology, jugs, seises, pickups, phones. People tend to call the special type used offshore, *hydrophones*.

The electrical cable connecting the geophones to the recording instruments is called an *instrument cable*, or just a cable. The type used offshore is called a *streamer*, or a cable.

The magnetic sound recording tape is just called *magnetic tape*, or tape. The one on which the data is originally recorded is a field tape. Then there are intermediate tapes, and the final tape, from which the paper display is made.

The paper display of the data is a *record section*, seismic section, or just a section.

The wiggly vertical lines of information on the record section are *traces*.

The echoes from underground are called *reflections*. The alignment of wiggles across a record section, that represents a reflection, is also called a reflection. A *horizon* is the surface the reflection bounces from, also the reflection itself, generally one that extends over a fairly large area.

The time taken by the sound in going down to a horizon and reflecting back to the geophones is the *reflection time*.

Two terms have been adopted from signal theory. *Source* is used as a general term for any initiator of seismic energy, or group of them used together. *Receiver* is any device, or group of them used together, to detect seismic energy.

These terms are just a few to get started. Others will be taken up with the specific subjects.

Echo-Ranging

Seismic exploration is an echo-ranging system. That is, it includes both transmitter and receiver, and determines distance to objects by measuring elapsed time between sending energy out and receiving it, bounced back from the object. Note that earthquake seismograph is not echo-ranging, as it does not involve transmitting energy, but only an instrument to receive energy from other sources.

There are several echo-ranging systems that are better known.

The best known is radar. In radar, radio waves, traveling at the speed of light, the limiting speed of the universe, are emitted, bounced off an object, received, and the elapsed time measured. Its most common use is in marine navigation, from the vessel. At the speed of light, that is *some* short time that is measured. The signal is usually received by a directional, but rotating, antenna, and the result shown on a circular screen with the vessel's position represented by the center of the screen.

Bats have an echo-ranging system, of which two statements are commonly made—that they have a built-in radar, and that they had radar a million years before we did. Well, it isn't exactly radar, as it uses sound rather than radio waves, but it is used for navigation in much the way radar is. The bat emits vocal sounds, and hears the return. The bat's brain must then do some fantastic data processing to avoid an obstacle here, track an insect's flight there, all in the vicinity of other bats, issuing their own signals. And he does the processing in "real time," as he goes along.

The water depth recorder also uses emitted and reflected sound. It produces a strip chart of the water bottom along a boat's course.

Reflection of Sound

Sound is a longitudinal vibration of matter. That is, it is a series of compressions and decompressions expanding outward from its origin. Its velocity depends on how fast vibration can be propagated through the material it is traversing. So, the velocity depends on the substance, its physical state (whether solid, liquid, or gas), and to a lesser extent, on the temperature and pressure of that material. The "compressional wave", when it meets a different substance, sets it vibrating too, but at the different velocity of that substance. The vibration for this new material is propagated forward, but also, the vibration sets up a vibration in the first substance, expanding back through it. This is an echo, or reflection of the sound.

These reflections occur at any change in the velocity of propagation of the vibrations, but are stronger where the change of velocity is greater. The velocities that seismic exploration is concerned with are about 1,000 ft/sec in air, about 4,900 ft/sec in sea water, and from around 1,500 to 26,000 ft/sec in the earth. The 1,500 ft/sec is for soil at the surface of the earth, and the 26,000 ft/sec applies to some deeply buried, hard rocks. Rocks in the earth generally have higher velocities as they are more deeply buried, and harder rocks tend to transmit sound faster.

Sedimentary basins are composed of layers of different kinds of rock mostly deposited smoothly on one another. There is a "velocity interface," a change from one velocity of propagation to another, at each change from one kind of rock to another. There is a seismic reflection at each of these interfaces, with more of the sound reflected at the greater changes. Density of the rock enters into the amount of reflection also, but velocity contrasts are usually much greater, so people tend to speak of velocity interfaces without mentioning the density difference.

The expanding compressional waves are more or less spherical, so in sketching in cross section the behavior of seismic energy, arcs drawn with a compass illustrate the situation well (Fig. 1-2). However, it is also convenient to consider ray paths, i.e., lines representing the route sound took in going from source down to point of reflection, and up to geophone (Fig. 1-3). A ray path is the route taken by one point on the wave front (Fig. 1-4).

A basic geometrical fact for seismic exploration, is that a thing rebounding from a surface, leaves the surface exactly as steeply as it approached, or "the angle of incidence is equal to the angle of reflection" (Fig. 1-5). This applies to a tennis ball bouncing from a court, a light reflecting from a mirror, sound from a rock layer.

For sound, it can be illustrated best as a ray path. If the ground and a reflecting surface are both flat, and a sound is produced on the ground and received by a geophone some distance away, then the reflected path will be: source to reflection point to geophone—two straight lines with equal angles (Fig. 1-6). So the sound is reflected at a point on the

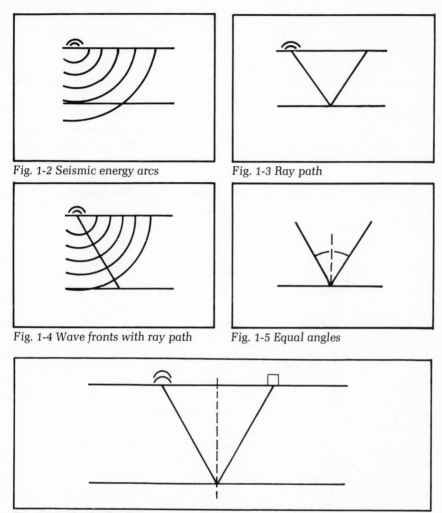

Fig. 1-2 Seismic energy arcs   Fig. 1-3 Ray path

Fig. 1-4 Wave fronts with ray path   Fig. 1-5 Equal angles

Fig. 1-6 Path of reflection

rock layer directly beneath the midpoint between sound source and geophone.

This is a basic principle of seismic exploration. A seismic shot yields data from midway between source and receiver.

This picture depends on the assumption that the ground and the rock layer are both flat. But what if that isn't true? Suppose the reflecting horizon dips. Consider first the situation when the source and receiver are at about the same place.

The angle of incidence and the angle of reflection must be equal. The only way this can happen is for the energy to strike the layer perpendicularly. This fits common sense knowledge too. To throw a ball and have it return to your hand, you must throw it so it strikes something at a right angle. To have it bounce back from the floor, you throw it straight down (Fig. 1-7). But if you throw at a leaning sheet of plywood, then you have to throw outward, so it hits perpendicularly to the plywood (Fig. 1-8).

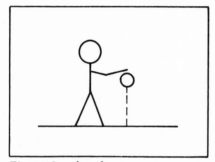

Fig. 1-7 *Level surface*

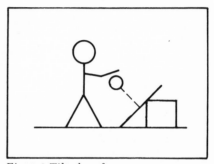

Fig. 1-8 *Tilted surface*

So in this case, the point of reflection of sound is not beneath the source and receiver, but offset somewhat. Which way? In the updip direction, up the slope of the tilted layer (Fig. 1-9).

Similarly, when the source and receiver are a distance apart, the point of reflection is moved updip. Again, for angle of incidence and angle of reflection to be equal, the reflected point is offset in the updip direction (Fig. 1-10).

However, most rock layers do not dip steeply, and the reflection point is not offset far from the midpoint. So, for practical purposes, the reflected point is usually considered to be midway between source and receiver.

The main value of seismic reflection in exploration is to provide information on relative depths of a layer, i.e., the reflection of a specific layer is recognized along a seismic section, and the reflected times determined at different points on the line. These times are clues to relative depths to the reflector. A smaller than normal time indicates a point where the reflector is structurally high.

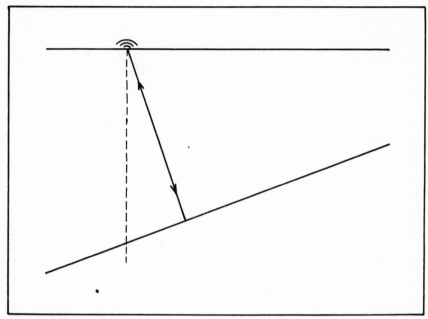

Fig. 1-9 *Offset with dip*

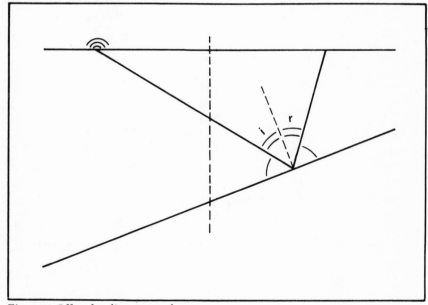

Fig. 1-10 *Offset for distant geophone*

Group

In describing what happens when a source and receiver are in certain positions, they were treated as single points. However, useful seismic data, which are reflected from deep in the earth, are received by a geophone along with considerable "noise," or unwanted energy. So it is worthwhile to arrange things so the wanted data is favored, while discriminating against the noise.

Sound that travels direct from the shot to the geophone, is almost horizontal for a distant geophone (Fig. 1-11). But the energy reflected from below is traveling more nearly vertically. It is desirable to favor the vertical energy, and diminish the effect of the horizontal.

For this purpose, the one geophone that would produce a trace, is replaced by a group of geophones, spread out over the ground (Fig. 1-12). The vertical energy strikes all of them at about the same time, but horizontally traveling sound strikes the geophone nearest the source, then progresses to the others. The geophones in the group are connected electrically by wires, so they feed data into the cable as though they were only one geophone. So the horizontal energy, reaching the different geophones of the group at different times, is transmitted through the cable out of phase, with some wiggles cancelling others. So the total horizontal energy transmitted is reduced. Vertical energy, arriving almost simultaneously, is transmitted more or less in phase, so wiggles reinforce each other.

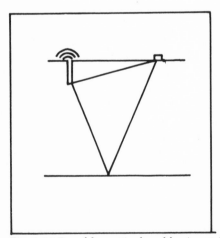 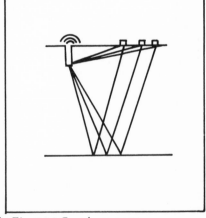

Fig. 1-11 Roughly vertical and horizontal   Fig. 1-12 Geophone group

Most seismic data is recorded from groups of geophones, rather than single phones. The geophones of a group are spread along the cable for a distance of several feet to several hundred feet.

Special types of geophone groups are sometimes designed, to perform better noise cancellation for a specific situation.

Groups are often "tapered," i.e., the geophones in the group are unevenly spaced, so more of them are near the center of the group, those farther out being fewer and farther apart (Fig. 1-13). This taper can be designed very specifically for reduction of the type of noise encountered in a particular area.

The noise-cancelling effect is exactly the same if, instead of several geophones, there are several sound sources. The sources can be used simultaneously, or one source several times, to be combined later on tape, to give the effect of several.

If there are several of each, source and receiver, the horizontal noise-cancelling effect is even greater.

Spread

The geophone groups are spaced along an electrical cable. There are usually twenty-four, forty-eight, or ninety-six groups on a cable. Each group puts one sequence of electrical signals, corresponding to vibrations of the ground, into the cable. The signals go through individual wires in the cable to the recording instruments, where a separate trace is produced from the information in each group.

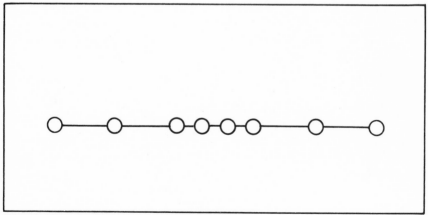

Fig. 1-13 Tapered group

The cables are usually from 1½ miles to more than 2 miles long, with the groups from 110 to 330 ft. apart. Sometimes the groups are longer than the spacing between them, so adjacent groups overlap. Although the overlap sounds paradoxical, the important thing is the placing of centers of groups. The horizontal extent of each group is just a noise-cancelling and data-averaging device.

The single-ended spread, or single ender, has the cable laid out in one direction from the shot point. Each geophone, or group, receives data from a point half way between it and the shot point (Fig. 1-14), so the whole spread gives information from the shot point half way to the farthest geophone.

After shooting a shot point, the cable can be left there, and another point shot at the other end of the spread. This gives information from the new shot point half way back (Fig. 1-15). So the whole distance between shot points is filled in. This is complete subsurface coverage, or single fold shooting, or 100% shooting.

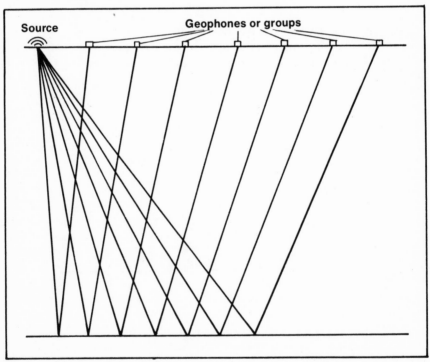

Fig. 1-14 Single-ended spread

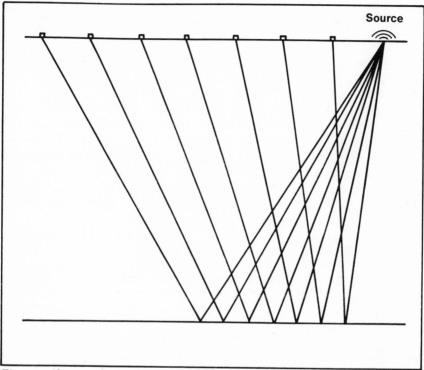

Fig. 1-15 Shot at other end

If a spread is laid out in two directions from a shot point, then the picture is just the same but with an additional spread going the other way. So subsurface data extends from the shot point half way to each end (Fig. 1-16). This is a split spread. 100% coverage is achieved the same way as before, with other points shot at the ends of the spread. However, when a point is shot at one end, notice that only half the cable is left where it is, and the other half moved around to extend the other way from the new shot point, so it has a split spread also (Fig. 1-17).

A string of shot points using either split or single ended spreads is a continuous line. It can be extended for any distance, giving complete subsurface information.

A buildup in coverage from 100% to higher percentages is achieved by the common depth point process, which covers the same distance more times, to produce better data.

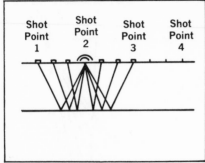

Fig. 1-16 Split spread

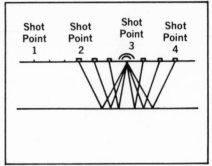

Fig. 1-17 Next spread on the line

Common Depth Point

At one time, there was only one data combining method available. This was called mixing, or, to make it sound a little more sophisticated, compositing. In its simplest form energy from two adjacent geophone groups was combined into one trace, just by hooking them to the same wire. Thus, the trace was an electrical composite of the data received at the two locations. This mixing could take place as described, out where the instruments were connected to a cable leading to the recording truck, or the blending could be done with the same effect in the truck, by connecting the wires that had brought the two separate sets of data to the truck. The first was called a ground mix and the second an instrument mix, or just a mix.

It worked fine for flat data, but tended to discriminate against dip as well as against noise. It did help in finding large, gently-dipping features.

A method of getting the advantages of combining data without smoothing it out so was invented. It is the Common Depth Point method. It is also called CDP, or referred to as stacking. CDP data come in percentages of stack—400%, 600%, etc.

The principle of CDP shooting is quite simple. A source and a receiver produce subsurface data from directly under the midpoint between the two (Fig. 1-18). Another source is placed farther from the center, and a receiver is placed an equal distance on the other side (Fig. 1-19). When energy from the second source is recorded at the second receiver, subsurface information will be from the same center point. So energy traveling by two paths gives information on the same point. Part of the longer path can be subtracted to make it the same length as the

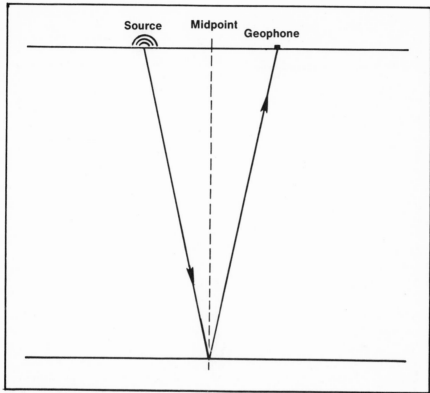

Source   Midpoint   Geophone

Fig. 1-18 Single source and receiver

other, and the two traces combined into one. Now the data enhancement has taken place without the smoothing of a mix.

That is what CDP, or stacking, is all about. Only 200%, or two-fold CDP was described, but it's easy to see how more routes of energy can be added to increase the multiplicity of stack (Fig. 1-20).

In actual practice, CDP shooting isn't so simple, and is more economical. Rather than choosing a spot, and taking a succession of farther shots received by farther instruments, the operation is made more wholesale. A long cable is laid out, with a lot of instruments along it. Then, shots are fired at fairly close intervals along that line (Fig. 1-21). The various energy paths are sorted out later in data processing for correction and combination of sets of paths that have the same depth point.

Offshore, a cable 1½ miles to over 2 miles long is trailed behind a

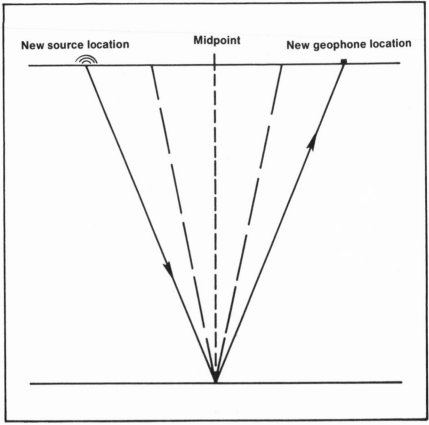

Fig. 1-19 *Two sources and receivers*

boat and sounds are produced near the boat. Again, the long and short energy paths through the same depth points are sorted out and put together in processing.

CDP includes some basic assumptions, that dictate shooting procedure. For one thing, the geometry applies to points strung out in a straight line. Only a slight deviation from a straight line can be tolerated. Fifteen (15) degrees is the generally-accepted limit to the amount of bend in a line permissible without being too damaging to the data. Where the line cannot be straight at all, the paths through the same or nearly the same depth points can be sorted out and combined. Or, for 3-D purposes, this may be done intentionally.

Another assumption is that there is no dip. Reflection takes place

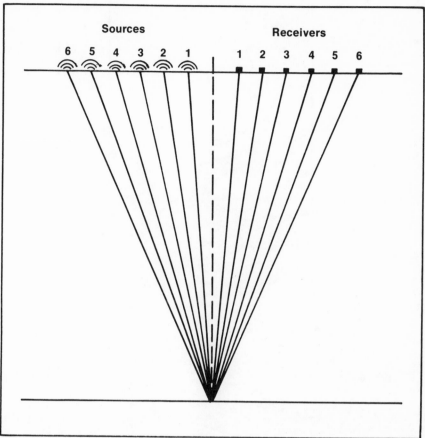

Sources   Receivers

6 5 4 3 2 1   1 2 3 4 5 6

Fig. 1-20 600% CDP

below the midpoint of source and receiver only when the reflecting horizon is flat. The shift away from the midpoint is not great for small amounts of dip though, so the method works well even for most dipping beds. Dip not only shifts the reflection point out from under the midpoint, but shifts differently for different energy path (Fig. 1-22). If source and receiver are close together, the shift is small. For greater source-receiver separation, the shift is greater. So the depth points are not common. However, the depth points are usually close enough that the "smear" (this type of poor stack) caused by the separation does less harm than failure to stack would cause. That is, the stacked data are still much better than 100% data.

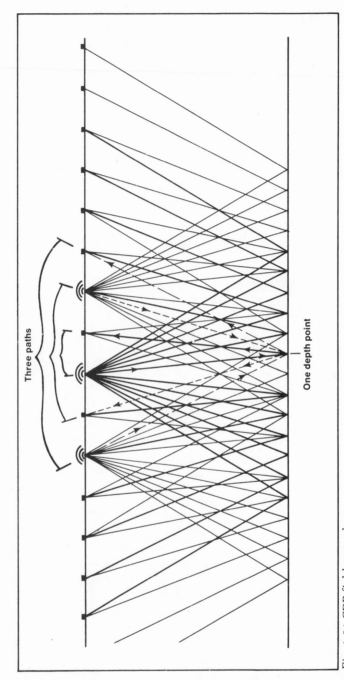

Fig. 1-21 CDP field procedure

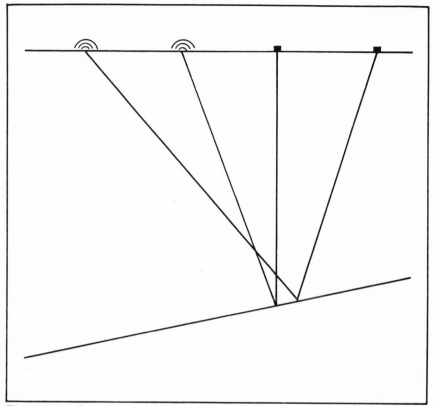

Fig. 1-22 Depth point not common

One of the main uses of CDP is in suppressing multiples, repeated reflections. This application is discussed under "Multiple Attenuation" in Chapter 5, Data Processing.

Refraction of Sound

When sound strikes a velocity interface, going from a substance that transmits it at one velocity to a different one with a different velocity, some of the sound is not reflected, but continues on its way. However, its path is bent at that point, so it goes in a different direction (Fig. 1-23). This change of direction is refraction. The amount of change depends on the degree of difference in velocity. No difference, no bend. Slight difference, slight bend. Large difference, greater bend.

Also, the bend depends on the angle at which the sound strikes the interface (Fig. 1-24). Perpendicular, no bend. Slight angle, slight bend, etc.

When the velocity interface is a change to a faster velocity, the bend is in the direction increasing the angle from the vertical, making the energy go farther away from the source. In the earth, most of the changes are of this type, since compaction tends to make deeper rocks denser and higher velocity. So most seismic refraction bends the sound away from the source, on the trip downward. Coming back up is the reverse (Fig. 1-25), with the path more nearly vertical as it approaches the top of the ground. The interfaces on the way up are mostly from fast to slower.

Downward traveling sound that meets an interface with a change to a slower rock, will be bent so it is more nearly vertical, (Fig. 1-26). So the sound is always more vertical in the slower material, less vertical in the faster.

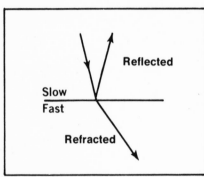

Fig. 1-23 Refraction

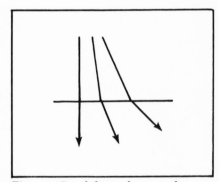

Fig. 1-24 Bend depends on angle

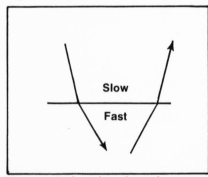

Fig. 1-25 Both paths refracted

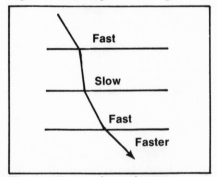

Fig. 1-26 Fast to slow refraction

The refraction of sound in rock is the same phenomenon as the refraction of light in substances that are transparent to it—air, water, glass—and that makes a stick standing in a pond look bent at the air-water interface (Fig. 1-27).

For any slow to fast rock interface, the sound arriving at an angle leaves at a greater angle. There is then some angle of arrival for which the continuing sound in the faster rock goes along parallel to the interface (Fig. 1-28). This is the "critical angle". Beyond it, reflection does not take place, but sound going along the interface vibrating the rock as it goes, will send energy up to the surface, from every point along the way. This is also a quickest path, as sound traveling down at the critical angle, along the higher velocity rock, and up at the same critical angle, will arrive sooner than sound that took a direct route through the slower material (Fig. 1-29).

Refraction plays a part in all seismic exploration. In reflection of sound, the paths are straight lines only in the oversimplified situation of going through rock of all the same velocity. Where different layers of increasing velocities are encountered, there are bends away from the vertical, (Fig. 1-30). If there are not discrete beds, but a progressive increase of velocity with compaction, as in a massive shale, the path of the sound is curved (Fig. 1-31).

If geophones are placed beyond the reflections, where the critical angle causes the returned energy to use a path refracted through a high speed layer, then refraction is used to determine things about the earth.

Fig. 1-27 Refraction of light

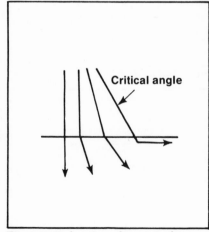

Fig. 1-28 Refraction along interface

This being beyond the reflections, is relative to the depth of the high speed layer, so geophones that are properly placed to receive reflected energy from deep horizons, will receive refracted energy from very shallow beds (Fig. 1-32). This is used in reflection exploration to correct for near surface irregularities.

Geophones placed much farther away, like several miles from the source, are at a refraction distance for deep horizons. This is used in the refraction method of seismic exploration. Refraction exploration, once the only form of seismic exploration, has been replaced by reflection so thoroughly that refraction is only used for some special situations. A brief chapter later in this book, Chapter 9, is devoted to refraction exploration, but otherwise, where seismic exploration is mentioned, it will refer to reflection exploration.

Records and Sections

Record sections are the main feature of seismic exploration. Field work and data processing are aimed at producing sections. And interpretation is just working on the sections to find oil.

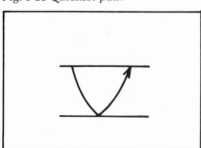

Fig. 1-29 Quickest path

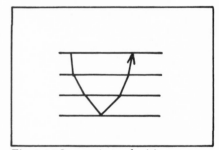

Fig. 1-30 Increasing velocities

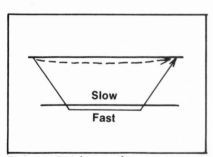

Fig. 1-31 Steadily increasing velocity

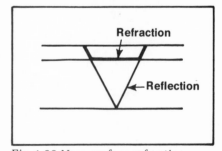

Fig. 1-32 Near-surface refraction

Sections are made up of traces. A trace is the data from one shot that is received by one geophone group, displayed as a wiggly line, or in some other form. In its conventional form, the data is represented as the wiggles on the line.

From early earthquake seismographs, came the word "trace"—the line was traced as the ground shook—and from early seismic exploitation, the terminology of the wiggles. If the trace is horizontal, with its start at the left and later times to the right, the wiggles will be up and down (Fig. 1-33). The up ones are therefore called peaks, and the down ones troughs. The convention is retained even when the trace is held vertically (Fig. 1-34). Then the peaks are to the right and the troughs to the left, but they are still called peaks and troughs. A trace with no wiggles at all, usually because it resulted from something being disconnected somehow, is a dead trace.

As the trace is a record of ground motion, a peak represents an upward motion of the ground, a trough a downward motion—or can be the other way around. Direction of traces is not thoroughly standardized, so traces from different operations may differ in this way, being said to have opposite polarities.

A seismic record is a display of all the traces from one shot (Fig. 1-35). It was for many years, until tape recording and record sections were developed, the only form in which seismic traces were displayed, and all seismic interpretation was performed on these unmodified records. The chief present day use of records is as monitors during recording, to spot check recording quality.

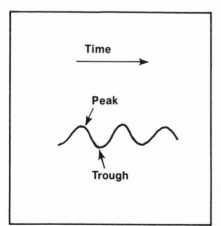

Fig. 1-33 Terminology of wiggles

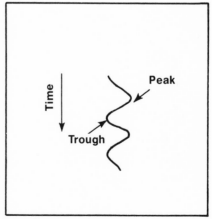

Fig. 1-34 Vertical trace

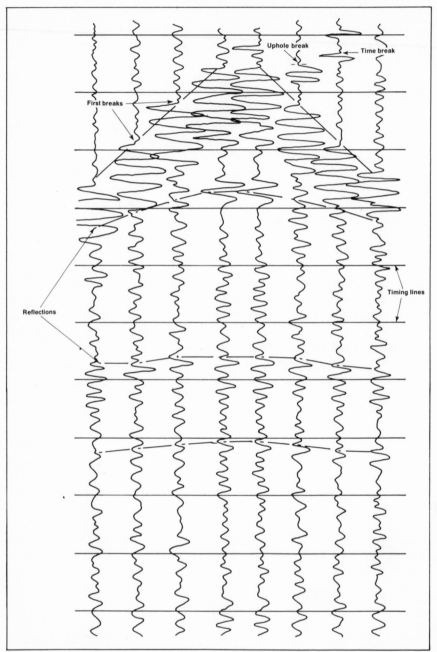

Fig. 1-35 Seismic record

The record may be 4 to 10 in. wide and 2 to 3 ft. long. There are usually 24 or 48 traces on it side by side, the same as the number of geophone groups on the cable used to pick up the data. The traces are so close together that peaks and troughs overlap other traces.

There are timing lines running across the record, to mark time elapsed, usually one line every 10 milliseconds.

Progressing down the record, from early to later times, the traces are at first "quiet", i.e., nearly straight, with only small wiggles, caused by wind or other slight disturbances.

Then, the first event on the record is the "time break". This is an abrupt burst of energy on one trace, appearing as a sharp, strong wiggle. It is made at the instant the shot is fired or other source initiated, and is used as a starting point for all events on the record. The timing lines after the time break are counted, with interpolation, to determine the time at which any event occurred. The time break does not particularly belong to any one geophone group more than another, so it is arbitrarily put on a trace that allows it to be out of the way of other events.

On records taken by firing an explosive in a hole, the next event is the uphole break. It is recorded by a geophone on the ground, very near the hole, maybe 10 ft. away. (Fig. 1-36). The first sound to reach this geophone traveled directly to it from the shot, and is called the uphole time. The geophone is called the uphole jug, shot point seisometer, or SPS. On records taken on land with surface sources (no shot hole), and on offshore records, there is no uphole time.

Next on the record come the "first breaks". They are the first wiggle on each trace, as the sound reaches the geophone groups on the cable. The nearest of the groups to the shot point is from 100 to several hundred feet away, so the first breaks are all later than the uphole break. The first breaks from the further groups appear later and later on the record, so the line of first breaks slants away from the near trace (trace from the group nearest the source). If the spread extends in one direction from the shot, the first breaks will slant in that direction. If the spread is split, they will slant in two directions. The first break lineups normally are in straight lines, or segments of straight lines, indicating the velocities of the layer or layers near the surface, that the sound travels along. That is, the sound is refracted so it travels along the upper surfaces of the layers.

After the first breaks are the reflections. They appear as wiggles that line up, from trace to trace, across the record. The reflected energy goes a different route from the refracted first breaks, a route that makes the

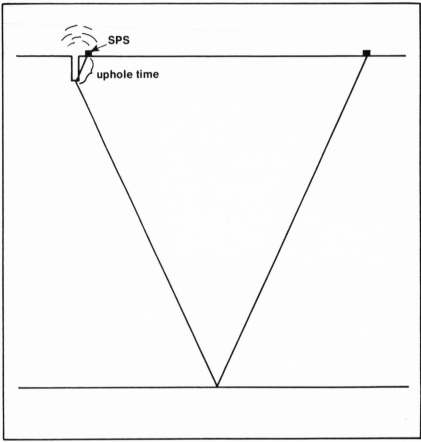

Fig. 1-36 Uphole time measurement

reflections be curved, bending downward with distance from the shot, rather than the straight line segments of the refracted energy. The shallowest reflections curve considerably, and the curvature becomes less for the deeper reflections. The curvature is called normal moveout. It is discussed more fully in the next chapter.

A large number of traces are combined to make a record section, originally a cross *section* made up of *records*, by just placing records side by side (Fig. 1-37). A modern record section, seismic section, or just section, is more subtly constructed, but has the data from a number of records adjusted and combined to make a display that looks strikingly like a cross section of part of the earth (Fig. 1-38).

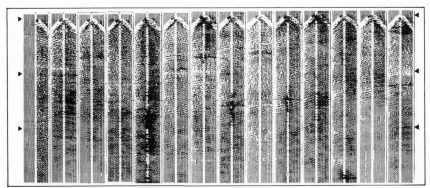

Fig. 1-37 Early record section   *Credit: Geoscience Technology Services*

The section has hundreds of traces side by side, overlapping. Shot point numbers are printed along the top of the section. They can be matched with the same numbers on a map, to locate the shot points of the section on the ground. There are timing lines, in the form of long straight lines across the section.

The reflections run across the section, but without the normal moveout curvature of reflections on records. The curvature is removed in the data processing that produces the sections.

The reflections on a section more or less show the underground rock layers in cross section. With limitations, the rock layers go up, down, are broken, etc., as the reflections indicate. The inexactness of depiction of the subsurface will be taken up in later parts of the book.

To use a record section, a person usually marks a reflection by drawing along it with a colored pencil. Then, interpolating between the timing lines, the times to the reflection at the shot points can be read. These times, plotted on a map, along with times from other sections, are contoured, showing the configuration of the rock layer, from which locations for drilling wells may be deduced.

Sections are discussed in more detail in Chapter 6, Forms of Sections, following description of data processing, which produces the sections. The interpretation of sections is discussed in Chapter 8, Interpretation.

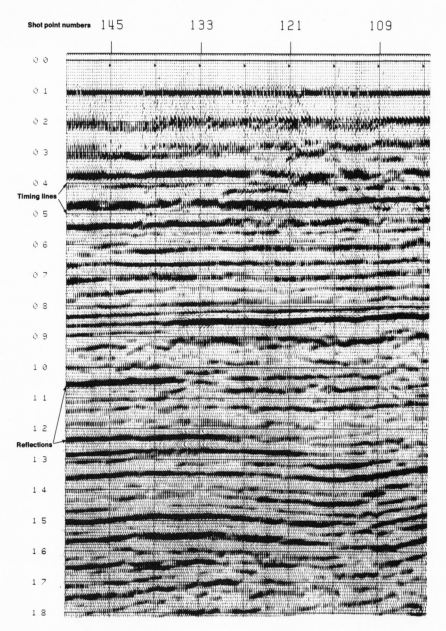

Fig. 1-38 Modern record section   Credit: Western Geophysical Co.

2

Phenomena

Seismic investigations deal primarily with reflections. Other phenomena add to the picture, or interfere, and must be taken into account in interpreting the sections. Some phenomena require correction, or attenuation, to make the reflections more interpretable.

Weathering

The term "weathering" has a meaning in geology and a meaning in geophysics, but the meanings are not the same. To a geologist, weathering is unconsolidated material that has been broken up by wind, water, maybe by plants. A geophysicist refers to the material at the surface that has a considerably slower velocity of sound than slightly deeper rocks. Often the change in velocity is abrupt, say from 2,000 ft/sec in the weathering to 6,000 ft/sec immediately below it. The base of the weathered layer is sometimes the water table—the top of ground water. Air has a velocity of about 1,000 ft/sec and water about 5,000 ft/sec, so the change from soil with air between the grains to that with water is the abrupt kind of change that can define the base of the seismic weathering. The water table can change with the season or with recent rains, etc. so seismic weathering in an area can change from month to month, or even from day to day.

Calling the seismic weathering the low velocity layer, avoids the confusion with geological weathering. However, there may be deeper layers with low velocity, so there is another possible confusion. Paul Lyons introduced the expression LVL (from "low velocity layer"). This expression pretty well avoids both sources of confusion, and is used in such expressions as "LVL crew", a group that drills shallow holes and shoots in them to determine thickness of the slow layer for an otherwise surface-source crew (vibrator, etc.). Most people still use the term "weathering" though, so, as one purpose of this book is to make familiar the expressions used in geophysics, we're stuck with "weathering".

The weathering is a very critical zone in seismic operations. Shots fired in the weathering tend to be low-frequency, as the relatively loose dirt doesn't transmit the higher frequencies well. The long, spread out wiggles on the record don't contain as precise information as higher frequencies, so it is preferable to fire shots below the weathered layer. Then the higher frequencies are transmitted better, and the uphole time, from shot to surface, includes all of the weathering, so the reflections can be corrected for the time spent in this slow material.

In diagrams, forms, notations, "Wx" is often used as a symbol for weathering.

Normal Moveout

On a field record, reflections are curved. The curvature is downward from the shot point, and is concave downward (Fig. 2-1). This curvature is called normal moveout, as it is the normal way reflections behave as the traces are further (move out) from the shot point, rather than representing dip or any other local phenomenon.

To see why reflections curve, consider the simple reflection situation of a spread in one direction from the shot point, and a single flat reflection. A geophone very near the source will receive energy that traveled almost straight down to the reflecting horizon and almost straight back up. A more distant geophone receives energy that went out at an angle to the reflecting horizon, and up at an angle to the geophone (Fig. 2-2a). So, even though the horizon was flat, the sound traveled a longer path, taking a long time to get to the geophone, so it appeared farther down the record, making it look deeper.

Now consider these two energy paths as though they were sticks, hinged at the reflection points. Unfold them and let them hang straight down, to compare their lengths. Suspend them from the midpoints between shot and geophone, so they will go through the depth points (Fig. 2-2b). Then they better represent traces of two way time, at the points they apply to in the subsurface. Looked at this way, of course the far trace is longer than the near one. If other traces are added at intermediate distances, their lower ends will form a curve (Figs. 2-3a & b). that is the normal moveout curve. Now drawing the same thing again, but for a deeper horizon, the curve will be similar, but flatter, as the geometry is different for the deeper level (Figs. 2-4a & b).

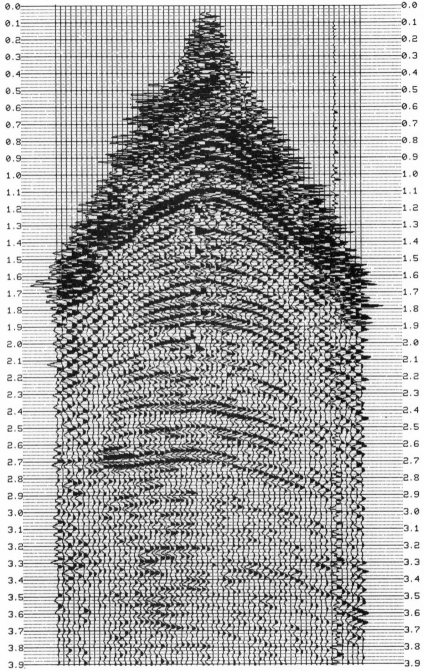

Fig. 2-1 Curvature of reflections   Credit: Houston Oil & Minerals Corp.

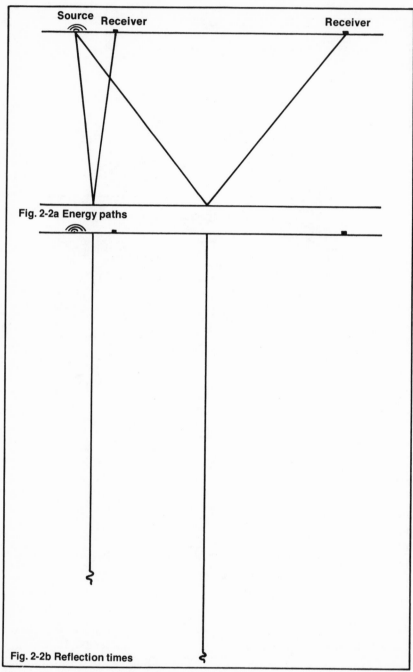

Fig. 2-2a Energy paths

Fig. 2-2b Reflection times

Fig 2-2 Two receiver distances

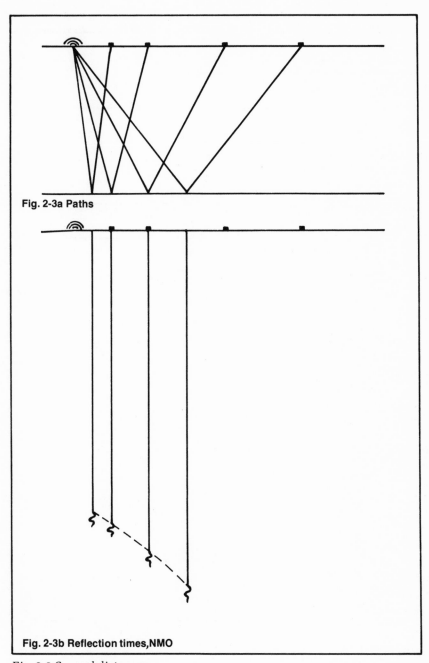

Fig. 2-3a Paths

Fig. 2-3b Reflection times,NMO

Fig. 2-3 Several distances

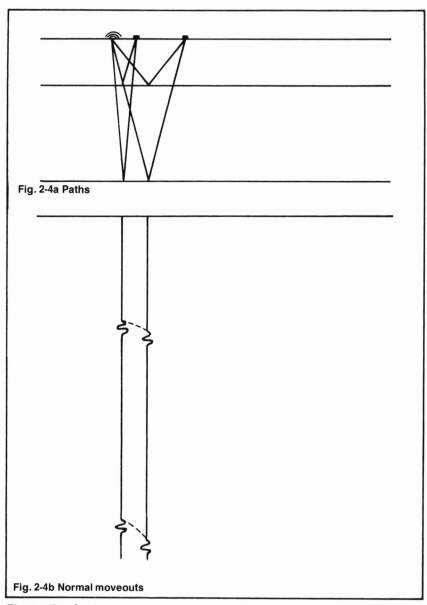

Fig. 2-4a Paths

Fig. 2-4b Normal moveouts

Fig. 2-4 Two horizons

The velocity of sound in the different kinds of rock enters into the picture too, as the far trace represents more travel in the shallow, slower, part of the section.

For brevity, normal moveout is referred to as NMO, or sometimes just moveout. The word moveout alone is not clear though, as any other curvature on the record, caused by dip, etc., is also moveout, although not normal moveout.

Multiples

Sound is reflected from places where there are sharp differences in the velocity of sound transmission. A seismic reflection on a section represents sound that goes down to a rock layer of different velocity, is reflected, and comes back up to the top of the ground (Fig. 2-5). However, the base of the weathering has a strong velocity contrast too. So, some of the energy should reasonably be expected to be reflected back downward. It is. And some of that can be reflected off the rock layer, up to the surface again (Fig. 2-6). This is a multiple reflection, or just a multiple. The geophones receive reflected energy twice from that same rock layer. The second one, the multiple, takes about twice as long as the first one, as it goes most of the same route, but twice. So, on the record section it appears to represent another rock layer, below the first one (Fig. 2-7). The section has, instead of information from two horizons, two sets of information from one horizon. The multiple's reflection time read from the section is about twice the time of the first

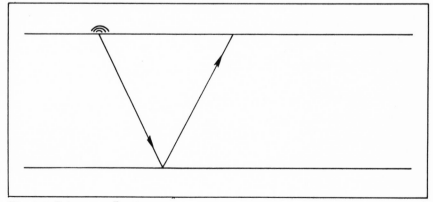

Fig. 2-5 Primary reflection

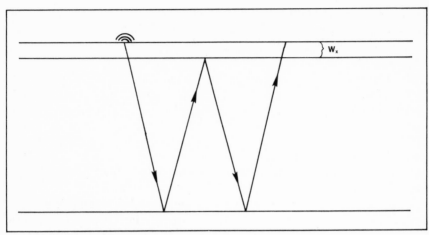

Fig. 2-6 Multiple reflection

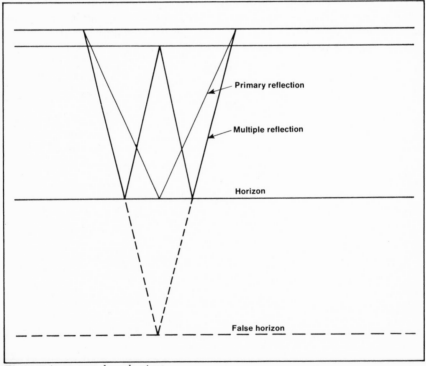

Fig. 2-7 Apparent deep horizon

reflection, the primary reflection or just primary. A measurement using dividers will therefore show it to be about twice as far down the record.

And, if there is enough seismic energy, what's to prevent another bounce from the surface, and another multiple? Nothing. The multiple will be at three times the time of the primary, as it traversed the same path three times (Fig. 2-8). And so on.

The successive multiples keep getting weaker and weaker, but will show as far as data enhancement can bring them out.

The multiple's time was given as *about* twice the primary's rather than exactly twice. This is because the primary goes from the shot to horizon, to geophones at the top of the ground, but the multiple's extra path may be from horizon to weathering to horizon.

The top of the ground also has a velocity contrast, and it, too, may sometimes reflect sound downward to produce a multiple.

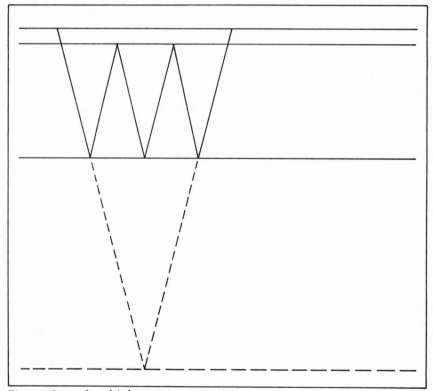

Fig. 2-8 Second multiple

Multiples can take several forms, depending on what layers they bounce between, and how many times they bounce. They diminish as they bounce back and forth, so there is a limit to how many times energy can be reflected and still be strong enough to be recorded. However, the enhancement of deeper reflections and weaker energy improves these multiples along with the wanted data.

Offshore, multiples are produced by reflecting surfaces at both top and bottom of the water. These can both be interfaces with strong velocity contrast. In this case, energy can bounce back and forth between these two surfaces a number of times (Fig. 2-9), so the section

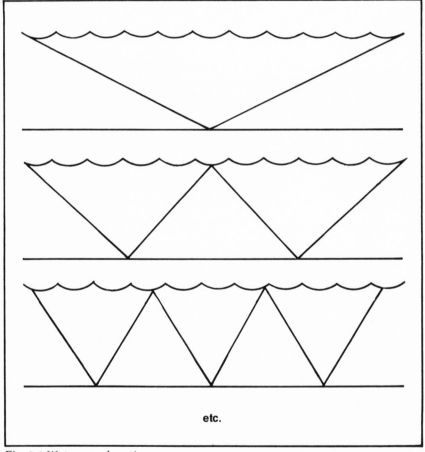

Fig. 2-9 Water reverberation

is dominated by multiples of the water bottom. This special multiple situation is referred to as a reverberation. A section with strong reverberations is called a "ringing" or "singing" section, and looks pretty much the same all the way down. The reflections seen are mostly just repetitions of the sea bed. In data processing, there are computer programs that fairly well take care of reverberations in most offshore data, but not in very deep water (say 1000 ft or so), Fig. 2-10.

Similar reverberations sometimes occur on land, bouncing between the top of the ground and the base of the weathered layer. They can be

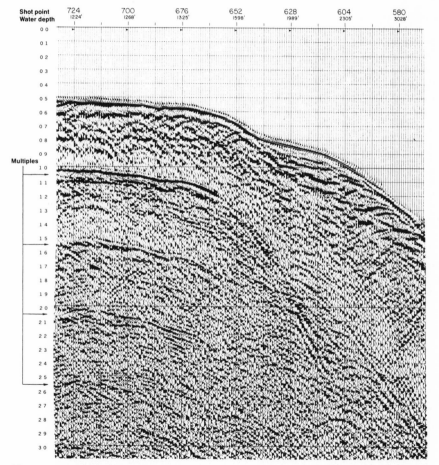

Fig. 2-10 Deep water multiples   Credit: Western Geophysical Co.

handled in the same way as water reverberations, but perhaps not as well, because shooting on land does not normally have the high degree of common depth point stack that is standard offshore. Also, the two surfaces on land are often more irregular than at sea.

Multiples can bounce back from any good reflecting surface. Energy may go down to a horizon, up to another rock layer, down to the first one, then up to the surface. This is an interbed multiple (Fig. 2-11), as it does its extra bouncing between two beds of rock.

A reflection that, either on its way down or on its way up, bounces between two layers, is a pegleg multiple (Fig. 2-12).

The interbed and pegleg multiples do not occur at twice the time of the primary, but at the time of the primary plus the time spent in the extra bounce. This is the general situation for all multiples, the primary plus the extra bounce or bounces.

In data processing, CDP stack combines traces after correcting for the

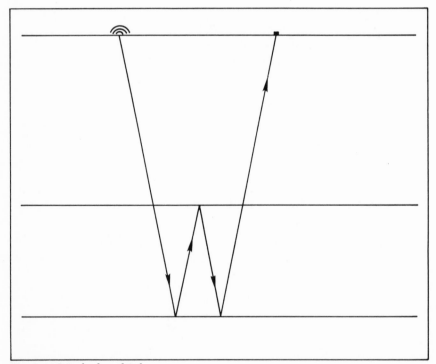

Fig. 2-11 Interbed multiple

normal moveout of primary reflections. This favors the primaries and weakens multiples. This is discussed in Chapter 5, Data Processing, under Multiple Attenuation.

Ghosting

A ghost is a special type of multiple.

When the source is an explosive in a hole, it is usually shot below the seismic weathered layer. Sound from the shot goes out in all directions. Some goes downward and is reflected back by some formation. It is recorded as a reflection. Some, of course, goes upward, and part of that may be reflected back down by the base of the weathering. Then it too can be reflected by the formation (Fig. 2-13). This is a ghost.

The shot usually isn't far below the weathering, so the ghost, reflected from the weathering, isn't far behind the more directly-reflected

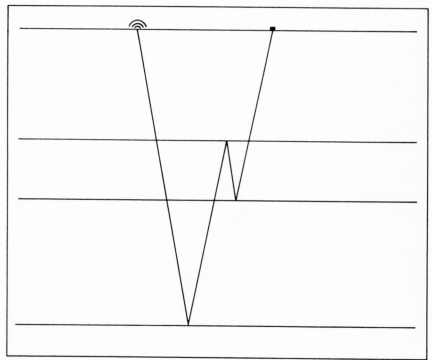

Fig. 2-12 Pegleg multiple

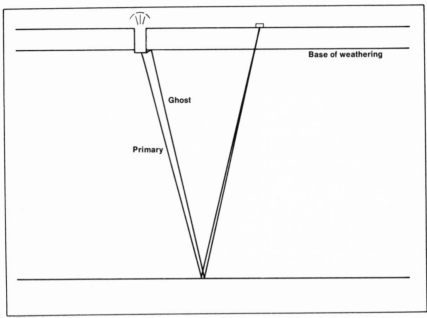

Fig. 2-13 Ghost

energy. It is often so close behind it that the two blend to make one
rather complex band of energy on the record (Fig. 2-14). This, in itself,
wouldn't be very harmful, but when the source is dynamite in a hole,
another shot at a different depth in the same hole, won't be the same
distance below the weathering. So the reflection and the ghost will
combine to make a different-looking band of energy (Fig. 2-15). And a
shot fired in another hole may also be different. So the character of the
records will change from shot point to shot point and from shot to shot
in the same hole, not with changing geology, but with vagaries of shot
depth. This can make it very difficult to determine what is going on in
the subsurface.

Ghosts may also occur when energy is bounced back from the surface
of the ground, rather than the base of the weathering (Fig. 2-16).

One partial answer to the ghosting problem is to shoot deep holes. If
the shot is far enough below the weathering, the reflection and the
ghost won't meet, so won't blend into a single band of energy. This
helps, but doesn't keep the ghost from interfering with some other
reflection.

A de-ghosting computer process has been developed. After this

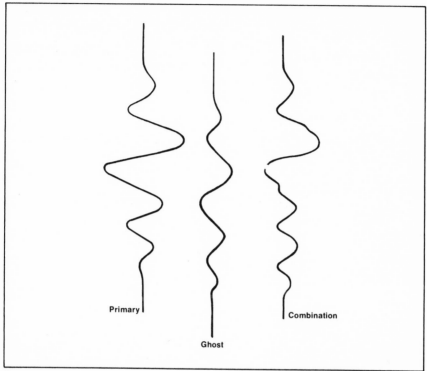

Fig. 2-14 *Effect of ghost on primary*

de-ghosting, another process, deconvolution, can help considerably in improving record quality and making records correlatable.

A ghost can also occur when a surface source is used. The extra path is from surface to base of weathering to top of ground (Fig. 2-17).

Bubbles

In offshore work, the most distinctively water-type problem is the formation of bubbles.

An explosion, any explosion, is just a very sudden expansion. Usually a solid or liquid becomes a gas, or a gas is released from confinement. In either of these cases, there is, suddenly, a rapidly-expanding gas.

When the explosion takes place underwater, the expanding bubble pushes water ahead of it. The water immediately ahead of the gas

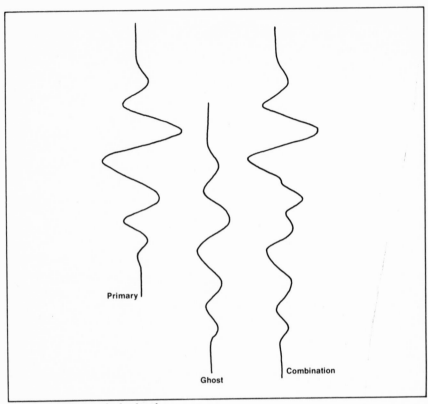

Fig. 2-15 Different hole depth

moves at a high speed (Fig. 2-18). After some short period, the force of the explosion is no longer great enough to move the water. But the fast-moving water can't stop immediately. Its momentum causes it to continue to move outward from the site of the explosion. When the momentum is used up, the bubble of gas is overextended, so the weight and pressure of water cause it to contract (Fig. 2-19). The leading water moves inward so rapidly that its momentum causes it to go too far, compressing the gas. The compressed bubble then expands again (Fig. 2-20). And so on, in diminishing expansions and contractions.

The second expansion is sudden, like the original explosion, although not as strong. To the seismic instruments and on the record it appears as though a second shot, not quite as strong, was fired. But of course, it isn't a convenient time for another shot. Reflections from the first one are still being recorded. So the two sets of reflections, from the

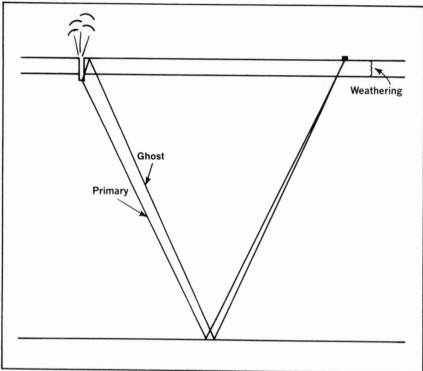

Fig. 2-16 Alternate ghost

first and second expansions, are all mixed together, And, third and later expansions are also mixed in.

In the various shooting methods used, there are several techniques to reduce the effect of the later expansions relative to the first one, both in special design at the source, and in processing.

Noise

The word "noise", in seismic exploration, is like the word "weed", in gardening. A weed isn't a certain kind of plant. It's just whatever grows that isn't wanted.

Similarly, seismic noise is the sound energy that isn't wanted. It even contains information, but not the kind wanted at the time. Generally, the wanted energy is the primary, one bounce, reflections, with all

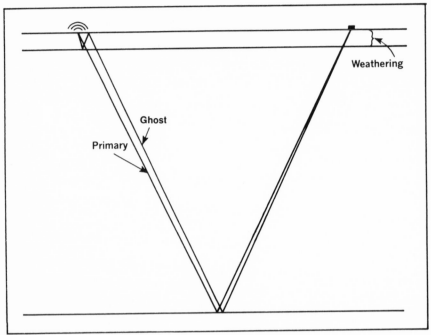

Fig. 2-17 Surface-source ghost

other energy considered noise. Effort is made to enhance signal, the information wanted, and weaken noise; to increase the signal to noise ratio, or S/N.

Kinds of noise encountered in seismic exploration include:

Multiples—repeated reflections;

Diffractions—curves formed at sharp breaks of reflecting horizons;

Random lineups on record sections;

On land

Outside noise—wind, cattle, etc.;

Hole noise—clods, etc., falling after a shot in a hole;

Ghosts—reflections from the base of the weathering, etc.;

Offshore

Bubbles;

Ocean waves;

Reverberations.

There are different techniques for treating these different types of noise, in field arrangements and in data processing. Filtering can reduce the high and low frequencies that are not from reflections.

Fig. 2-18 Underwater bubble

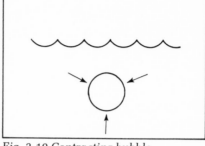

Fig. 2-19 Contracting bubble

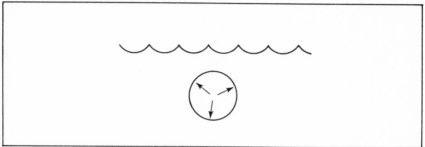

Fig. 2-20 Second expansion

Multiples are attenuated by common depth point stacking. Diffractions are diminished by migration.

Wavelet

An explosive seismic source produces a sudden, brief sound containing all frequencies. In going through the earth to a reflecting horizon and back to the surface, and then through the seismic instruments, it becomes stretched out into a "wavelet". A wavelet has one or two peaks and one or two troughs, and extends for about 50 to 100 milliseconds (Fig. 2-21). The peaks and troughs are of varying amplitudes, the highest occurring about 30 milliseconds after the start of the wavelet. The wavelet is the basic seismic response, the thing that would be recorded if there was only one reflecting horizon in the subsurface.

With just that one velocity interface, the wavelet recorded would have a polarity depending on which way the velocity change went, slow to fast or fast to slow, and an amplitude depending on the contrast between velocities (and densities).

In the real earth, there are a great many velocity interfaces, some with large contrast and many with small, some in one direction and some in the other. So the trace recorded is composed of a great number of these wavelets, some large and some small in amplitude, some turned one way and some the other, interfering with each other and blending together into a trace in which the individual wavelets are not separately visible (Fig. 2-22).

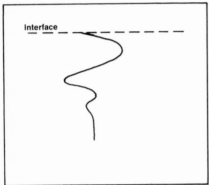

Fig. 2-21 Wavelet   Fig. 2-22 Trace, made up of wavelets

Another type of wavelet, a "zero-phase" wavelet, is symmetrical about its highest peak, which is at the time of origin of the wavelet, the velocity interface in the rock. This type of wavelet is the one applicable to vibrator data (Fig. 2-23). It can seem to start before its cause, because the section is made by a reconstruction of data, with whatever shifts are necessary.

Velocity Variation

Seismic velocities are subject to variations, both vertically and horizontally.

The velocity of sound in rock varies vertically with compaction of the rock, and with both vertical and lateral lithologic change. Compaction tends to increase the velocity with depth. The more overburden, the higher the velocity. This causes velocities generally to increase with depth. However, changes in lithology can go either way. A change from a shale to a carbonate, in the shallow parts of the subsurface, usually produces an increase in velocity, and from a carbonate to a shale a decrease. Compaction operates with duration of burial, weight of overburden, and compressibility of the rock, i.e., how long and how hard it was mashed, and how soft it was. A shale increases its velocity fairly rapidly with depth. A carbonate is more resistant, and retains about the same velocity regardless of depth. Salt is almost completely incompressible. So a salt dome will have a velocity of around 14,500 ft/sec all through its height, while the surrounding sediments will

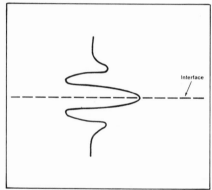

Fig. 2-23 Zero-phase wavelet

increase in velocity with depth, being slower than the salt at shallow depths, and faster in the deeper part of the section.

So in general, velocity increases with depth, but at an uneven rate, and with occasional decreases.

Horizontally, velocity changes with lithologic change and with dip. Thus if a sand grades into a shale, a velocity change can be expected. And if there is a lot of dip, then for a certain depth there will be a different formation and therefore a different velocity. Also the velocity of the formation will change with dip, as its depth of burial and therefore overburden, changes.

All these variations indicate that velocity is a highly variable factor in exploration. With steep dip or faulting, velocities obtained at a well may not apply a mile away from the well. On the other hand, a group of us once used velocities from a Louisiana well for an area near the Orinoco River in Venezuela. It worked pretty well.

Velocity is discussed at greater length elsewhere in the book. In

Chapter 5, Data Processing, obtaining velocities from seismic data is described, and in Chapter 7, Data Handling, other sources of velocity information and the use of velocity in producing synthetic seismograms. Velocity is also discussed in identifying reflections in Chapter 8, Interpretation.

3

Field Operations

Program Planning

A plan of lines to be shot, a seismic program, is made before the crew arrives in the field. It consists of drawing lines on a map, so coordinates of the lines can be determined and used to locate the lines on the ground.

The program may be a grid of lines, particularly in a large new area. It may be detail lines, filling in a prospective area of an earlier grid. It may be made to go through well locations, so well data can help identify formations on the seismic record sections, and seismic and well data be better worked together. It may be only a line or so to check the basic idea of a prospect found by surface geology or from well logs.

The program must fit the terrain and other surface features. It may be necessary, economic, or required, to shoot on roads, off roads, in valleys, on ridges, around certain farms, around lakes or towns. So planning for shooting on land calls for consideration of many factors. Air photos, land-use maps, ownership maps, pipeline locations, etc., may be useful in planning. And even with very good planning, the program will likely be modified by unexpected obstacles.

The program map is usually large scale, so locations can be determined from it. It shows wells, old shot points, benchmarks, towns, streams. The map, or a copy of it, will go to the surveyor on the crew, so he can use it to establish on the ground the locations to be shot.

Some suggestions for program planning are given in Chapter 12, General Considerations.

Field Geophysics

Seismic information is acquired by field crews. They may be almost anywhere on earth—on cultivated land, in a desert, at sea, in a forest, on a mountain, in a city. The climate may be tropical, temperate, or arctic. About the only limitation in looking for oil is that there should be a sedimentary basin, or a good chance of finding one.

So there is great variety in the conditions on field crews. A seismic crew may be on a rolling boat, reviewing the highlights of their last time in port, and living for the next time. A crew may live in a small town, drive trucks out each morning, and shoot along the edges of plowed fields. A crew may be hundreds of people, hacking a way through a jungle, carrying equipment, and setting it up frequently to shoot (Fig. 3-1).

Shooting on land and at sea are so radically different that the two will be described separately.

Field Crew

An arrangement is made to have a seismic crew do the shooting. It may be a company crew, that is, a part of the oil company wanting the

Fig. 3-1 Portable instruments in field   *Credit: Seiscom-Delta Inc.*

work done. Then the arrangement is an interdepartmental matter, of requesting the crew, having it available at the time, bookkeeping on its expenditures, etc.

In most cases though, a crew is hired from a seismic contractor. Seismic contracting companies range from large multinational companies to a single crew operated by the owner of the crew.

A normal crew consists of about 20 men and about half a dozen trucks or other vehicles. Where equipment must be carried by hand, the crew can include some hundreds of laborers (Fig. 3-2).

Hiring a crew may involve requesting bids from several contractors, or may be a matter of getting a crew that is available in the area. A contract specifying equipment, supervision, fees, etc. may be signed. Often a client and a contract crew find they work well together, and the crew, after doing the initial work in an area, may be hired each time there is more work to be done.

The crew is given the assignment map. The client may plan to have a representative in the field when shooting begins, to see what record

Fig. 3-2 Crew on jungle trail   Credit: Seiscom-Delta Inc.

quality is obtained, and what problems there may be. He may also visit from time to time during the progress of the shooting. Some clients may put a representative on the crew for the entire period of the work.

The contract company usually has a field supervisor, who is on the crew at the start, and makes frequent visits during the shooting. He is the main liaison with the client while the work is going on. He is normally of higher rank in his company than anyone on the crew.

Permitting

In the United States, where most land, and the minerals under it, is owned by individuals, permission must be obtained from the owner in order to shoot on his land. In many areas, a fee of so much per hole drilled, or per mile, is paid for the permission.

So before shooting can begin, a permit man (if it takes all of a man's time) or the surveyor (if it can be done between surveying duties) calls on the owner, describes where lines are planned, and asks if they can be shot. Most owners agree, as the discovery of oil on their land can be profitable to them as well as to the oil company. They may also give conditions for the shooting, like going along the edge of a field, keeping gates closed or open, etc., and may give useful information on how to get in to places, where water can be picked up for drilling, etc. (Fig. 3-3).

After the crew has shot on the land, the permit man may make a second visit to the owner, to see if there are complaints, and pay for unusual damage to the land, like ruts, where a truck was stuck.

Surveying

A surveyor, using the locations on the program map, finds where those points are on the ground, marks them for other members of the crew to find, marks routes through gates, etc. to those points, and surveys their exact locations and elevations.

The survey instrument, transit or theodolite, is a telescope with a means for measuring angles horizontally and vertically (Fig. 3-4). It is mounted on a tripod, and is used to sight on a "rod", a flat board with markings on it, designed to be visible from a distance, and to indicate height above the base of the rod. Or an alidade may be used. It is not mounted on the tripod, but can be moved about on a "plane table," allowing a map to be made as the survey progresses (Fig. 3-5).

The survey is run from a known location and elevation, a govern-

Fig. 3-3

J. C. Knight

ment bench mark, abandoned well, etc., by the surveyor and his assistant, the rodman, taking turns moving forward, always one at the last surveyed point and the other at an unsurveyed location. As a check on the surveying, the survey is normally run in "loops", that is, extended around to meet earlier locations to see if the new location and elevation are the same as the old, or, rather, that the "mis-tie" is within acceptable limits.

The surveyor and rodman place stakes, flagging, metal tags, whatever shows up and holds up well in the area, at shot points and

Fig. 3-4 Surveyor with transit   Credit: Digicon Inc.

geophone group locations, to mark those points for the other units of the crew.

Surveying is a rather solitary occupation. The surveyor is usually ahead of the rest of the crew, with only the rodman accompanying him. And the rodman is, much of the time, just a small figure seen through the instrument telescope. So the surveyor tends to "talk to him" while sighting. "Get out from behind that tree." "Hold the rod straight, you're leaning."

He also has to rely on his own abilities, like the surveyor in North Dakota, who told me he'd never had his pickup stuck in the snow, just slowed down. It could always go as fast as he could shovel a path.

Energy Source

There are several ways of initiating seismic energy. An explosive can be detonated in a hole in the ground. A heavy weight can be dropped. Propane can be exploded in a chamber pressed against the ground. A controlled vibration can be imparted to the earth, etc.

In the case of an explosive like dynamite, one shot produces sufficient energy for seismic recording, although several shots in different

Fig. 3-5 Alidade, plane table, rod
Credit: Western Geophysical Co.

holes may be fired simultaneously for other reasons. The other sources do not supply so much energy individually, so they are regularly added together in groups, either by using several simultaneously, or by combining several later in data processing.

Shot Hole

When a charge is to be fired in a hole, the hole is from 20 to more than 200 ft. deep, usually in the range of 60 to 100 ft.

A rotary drill, like a miniature oil well drill, is usually mounted on a truck or other conveyance.

The drilling is done by a bit on the bottom end of an upright piece of drill stem, heavy steel pipe threaded on both ends. The drill stem is rotated, so the bit will break up the rock. A fluid, mud, water, or air, is pumped down through the inside of the drill stem, out through holes in the bit, and up around the drill stem, carrying the "cuttings", chips of rock, up with it. More lengths of drill stem are added at the top as needed to deepen the hole.

The drill has a "mast", to hold the drill stem upright, a "rotary table" to turn it, a pump or air compressor for the fluid, a winch to pull the drill stem out after the hole is drilled.

A truck-mounted drill travels with its mast horizontal, and raises it to the vertical position for drilling. It also has a "pulldown", to put the weight of the back of the truck on the drill stem to add more force for drilling. A truck-mounted drill that uses water or mud, carries a metal slush pit, with two compartments. The fluid coming out of the hole goes into it, the cuttings settle in the first compartment and can be shoveled out. The mud or water goes into the second, and is used again for drilling (Fig. 3-6).

An air drill just keeps pumping air down the hole, and letting it blow the cuttings out.

A driller usually has one helper, who can also drive a water truck to keep the drill supplied with water.

In special situations of easy drilling and difficult access, other types of drill may be used. A "pumper" drill, little more than a water pump and some light drill stem, can be hauled down jungle trails. No bit is

Fig. 3-6 Drill on swamp buggy   Credit: GeoSpace Corp.

used. Water is just pumped down the drill stem to wash soft earth out (Fig. 3-7). Or in areas of loose dune sand, a "puffer" drill with an air compressor, will similarly blow air down the drill stem, blowing sand out of the way.

After the hole is drilled, it is usually loaded with explosive by the drillers, for use later, when the recording people are at that shot point. Cylindrical units of explosive are fastened, end to end, to make up the size charge needed. An electric blasting cap is attached to the charge. The charge is put down the hole with poles, or lowered by the wire on the cap. Precautions are taken to keep other people from pulling it up. The upper end of the cap wire is secured at arm's length down the hole. An "anchor" may hold the charge down.

Later, to initiate seismic energy, the cap wire is attached to recording equipment, and fired by a signal from the equipment. Mud, rocks, cap wire may spray into the air in a brief fountain (Fig. 3-8).

Fig. 3-7 Pumper drill. Man on top is the "pulldown"
Credit: Seiscom-Delta Inc.

Fig. 3-8 Dynamite shot   Credit: Geophysical Service Inc.   Photo by J. Funk

After use, the hole is plugged a little way down, and filled in.

Dynamite is the most compact seismic source, and is used where supplies and equipment must be hauled on foot down jungle trails, up mountains, other such areas of difficult access for vehicles. In other areas, where vehicles can carry supplies, bulkier but less sensitive explosives consisting largely of ammonium nitrate are used.

Pattern Holes

The discrimination against horizontally traveling noise that is achieved by a geophone group spread out over a horizontal distance of a hundred feet or so, can also be obtained by firing a number of explosive charges over a horizontal distance.

Several shot holes are drilled in a pattern, and shots fired in them simultaneously (Fig. 3-9). The holes are usually shallower than a single hole would be, and the charges smaller than would be fired in a single hole. The shallower holes may total about the same number of feet of drilling as the single hole, and may have, divided among them; about the same amount of explosive. The shallower drilling may be easier, if it does not reach as hard a formation, but it requires more times of

Fig. 3-9 Pattern shot   *Credit: Digicon Inc.*

setting up the drill. The method is applicable to portable operations, where a light-weight, light-duty drill can be carried more readily. A disadvantage is that the shots may not be below the weathered layer.

Surface Explosives

In some areas where portability is critical, shot holes may be dispensed with, so it is not necessary to transport a drill. Then small charges are fired at or somewhat above the surface. The charges are put on poles a few feet above the ground or laid on the ground. They can be placed in a pattern, like pattern holes. The elevated shots, fired simultaneously, will strike the ground somewhat as a flat force. This system also depends on having an area that is isolated, so the sound of the explosion does not disturb people.

Explosive cord is another application of explosives not requiring a shot hole. An explosive cord is payed out and buried a few inches deep by a plow on a tractor. A special advantage of this method is that the charge is spread out horizontally to give a noise-cancelling effect like that of pattern holes or a geophone group. For use in forests and other areas of fire hazard, a special cord with a fire retardant outer layer has been developed.

Gas Gun

A gas gun for use on land is a chamber that can be pressed against the earth, with a quantity of combustible gas, usually propane, mixed with oxygen, or with oxygen-enriched air, detonated in the chamber (Fig. 3-10).

The gun is usually mounted on the underside of a truck, so the weight of the truck will hold it firmly against the earth. There may be a device to pull the chamber up, right after it is fired, so it will not strike the earth a second time. There is a vent to allow the gases to escape after the shot is fired.

In operating in the field, there may be several vehicles that fire simultaneously, advance a short distance, fire again. A number of these "pops" may later be combined to be treated as one shot.

Gas guns can also be made small enough to be carried by helicopter, and set down where needed for firing. This type has been used particularly in the arctic.

Weight Drop

The weight drop seismic source uses a heavy weight, lifted about ten feet, then dropped on the ground (Fig. 3-11).

The weight is flat, and is dropped so as to land flat. Immediately after the weight hits, hoisting it back up begins. This raises it as it bounces, so it won't strike the earth a second time, and create a second set of seismic energy, with a second set of reflections. Also, it is then shortly ready for the next drop.

Fig. 3-10 Dinoseis* gas guns   Credit: Petty-Ray Geophysical
                                                        Division, Geosource Inc.

*Dinoseis is a registered trademark and service mark of the Atlantic Richfield Corporation.

Fig. 3-11 Thumper* weight drop unit   *Credit: Petty Ray Geophysical Division, Geosource Inc.*

Weight drops are used largely in desert areas, where many sources of energy, over a large area, may be needed to get usable seismic data. Dry, loose sand and soil transmit sound very poorly. It is also an advantage to not have to transport an explosive to remote areas.

A weight drop truck will usually have the weight at the back. Light chains dangle around the weight, to warn people from walking under it. The weight itself may be a piece of battleship armor plate.

Vibrator

A vibrator produces seismic energy by vibrating a weight, with the weight on a pad that is held in contact with the ground, so the vibration is propagated into the earth.

The vibrator is usually mounted on a truck (Fig. 3-12), and some of the weight of the truck can be used to hold the pad in firm contact with the ground (Fig. 3-13). The vibrating weight may typically be around two tons.

The vibrator goes through a "sweep", from high frequencies to low, or low to high, in the course of, normally, seven to 15 seconds (Fig. 3-14). Later, in processing, the record is treated so it is as though a single shot had been fired.

The vibrator causes very little disturbance to the surroundings, so it is often used in sensitive locations, where another source might cause

*Thumper is a registered trademark and service mark of Geosource, Inc.

Fig. 3-12 Vibroseis* vibrator on marsh buggy   Credit: Western Geophysical Co.

Fig. 3-13 Vibrator vibrating in Wyoming   Credit: Geophysical Service Inc.

Photo by J. Funk

*Vibroseis is a registered trademark of the Continental Oil Co.

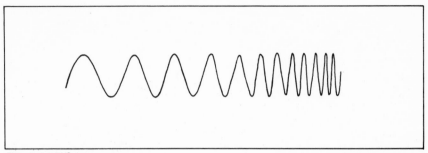

Fig. 3-14 Vibrator sweep

damage, like along highways or in cities. It has even been used to record in a tunnel through a mountain.

LVL Crew

When surface sources are used, a special small crew, an "LVL crew", may be used to obtain weathering information to supplement the data from the surface source. The LVL crew drills shallow holes and fires small charges in them, to record refraction data, from which base of weathering can be calculated.

Cable

Geophones are spread out in a line, connected by an electrical cable to the recording equipment. The cable is factory-made, maybe a half inch thick, and has as many pairs of wires as there are geophone groups, 24, 48, or more.

The cable is laid out, where feasible, by a truck with a cable squirter, that spews the cable from a pile on the truck, while the truck drives slowly down the line. After shooting, the squirter piles the cable back in the truck. Cable can also be carried wound on a "breast reel" worn by a man. He pays it out, or winds it up by a hand crank, as he walks along.

At intervals, the cable has "takeouts", pairs of wires to connect to the geophones of a group.

The phones that make up one group are pre-connected to each other by wire. They are placed in contact with the ground. The wire for the group is of a length to allow the geophones in the group to be placed a predetermined distance apart.

When all geophones are in position, groups connected to the main cable, and the cable connected to the recording equipment, an instrument is used to determine that all the groups are connected, none in reverse, no breaks in the cable. Then, or after corrections or repairs have been made, the spread is ready to be shot.

Communication between observer in the recording doghouse, shooter, and maybe jug hustlers when looking for cable breaks or other problems on the line, is by telephone, using wires in the cable or special telephone lines laid out when setting up. Or in some situations it may be by radio.

As shooting proceeds, there is a continual change of cable, segments from behind being brought around and connected at the end ahead of the shooting. This moving around is particularly characteristic of common depth point shooting, which is therefore called, from a field man's point of view, "roll-along."

Geophone

A geophone is a device for detecting sounds from the earth—geo, earth; phone, sound.

The type of geophone generally used in seismic exploration on land consists of a strong permanent magnet and a coil of wire, around the magnet, all in a rugged case (Fig. 3-15). The magnet is attached rigidly to the case, which can also retain the magnetic field. The coil, made of

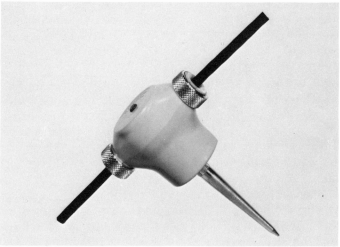

Fig. 3-15 Geophone with spike   *Credit: GeoSpace Corp.*

fine copper wire wound many times around, is connected to the case by a spring. The geophone is placed in firm contact with the ground. Then any shaking of the ground will shake the case and magnet. The coil, suspended on its spring, will not move as quickly. So the magnet moves up and down past the coil. Relative movement between magnet and coil generates electric current, on this small scale, just as it does in the dynamoes that power cities. The ends of the coil are connected to a "pigtail", a pair of wires extending from the case, to be connected, through the cable, to the recording equipment. The bits of current generated by the movements of the magnet make up the signal from the geophone. They are in proportion to the speed the magnet moves past the coil, so this type of detector is called a velocity geophone.

Geophones are set out by recording helpers, "jug hustlers". The "string" of geophones for a group is put on the ground at a set distance from each other. Good contact with the ground is assured. Geophones often have spikes on the bottom, so they can be stabbed into the ground. For frozen ground, they may be flat-bottomed, and need to be pressed into place, or grass scraped aside. Assuring the good contact is "planting" the geophones. A good plant is one in which the contact will allow vibration to be transmitted well from the earth to the geophone.

Recording

The purpose of the seismic source is to provide sound that will go into the earth, be reflected from rock layers, and be recorded for later enhancement and study. The geophones convert some of the sound into electrical energy, which the cable conveys to the recording equipment.

The recording equipment is a big, high-quality tape recorder. It can be mounted in a truck (Fig. 3-16) or on a marsh buggy, or break down into units that can be carried by people (Fig. 3-17).

Fig. 3-16   NEW INSTRUMENTS   J. A. Thomas

Fig. 3-17 Transporting instruments   *Credit: Geophysical Service Inc.*
Photo by J. Funk

The equipment is usually connected by wire, sometimes by radio, with the source, so a signal from the recording equipment turns on the source, firing the charge, dropping the weight, etc.

The equipment has computer-like panels with the usual flashing lights, numbers, dials, knobs. On one panel are two tape reels, with tape going from one to the other by way of a number of tension rollers (Fig. 3-18).

People watch the dials and make sure things are working correctly. When a shot, pop, etc. is fired, the tape reels turn for a few seconds, recording the data. Then the tape stops until the next shot. With an explosive in a hole, the next shot does not occur until the shooter has gone to the next hole, connected the cap wire, made sure no one is at the hole. With other sources, the operation is more continuous, with energy initiated maybe every few seconds.

Forms are filled out as a log of times of shots, problems encountered, etc. At intervals, a record is played back from the tape onto paper, for a spot check for problems. The paper record is developed automatically, either electrostatically or on light-developed paper.

After a number of shots, the takeup reel is full of tape, a new reel of unrecorded tape is put on, and the empty reel put in the takeup position.

The recorded reel is placed in a holder and sealed against dust, moisture, etc. It is packed with others for shipment to the processing center, at the next convenient time.

Fig. 3-18 Tape unit   *Credit: Geophysical Service Inc.*
Photo by J. Funk

Offshore

An offshore crew operates in a very different way from a crew on land. Rather than the units operating separately, the crew is all together on one boat. There aren't the discomforts of plodding through swamps, jungles, and snowstorms. But the boat may roll constantly. And the operation goes on 24 hours a day, so repairs and maintenance have to be done between lines, or while shooting. The shooting goes much faster and greater quantities of data are collected.

Offshore, leases, concessions, or production sharing contract areas are obtained from governments and permission to shoot is included in the arrangement. In addition, specific programs may require approval before shooting. A government representative may ride the boat through the course of the program.

Vessel

A seismic boat for normal offshore exploration is from about 150 to 200 or more ft. long. It is equipped to stay at sea, working, for a month to several months without supplies from land (Fig. 3-19). The crew consists of a ship's crew and a seismic crew. Both crews are doubly staffed, to keep the operation going day and night. The crews usually alternate in six hour shifts, with the breaks at meal times, so the people

Fig. 3-19 Seismic boat   Credit: *Petty-Ray*
Geophysical Division, Geosource Inc.

going on duty can eat before they start, and those going off can eat after work. Food is usually excellent and plentiful.

The boat is usually somewhat cramped, with equipment of one sort or other wherever there is room. And new temporarily needed gear is often added, and rarely removed, so the boat tends to get more cramped. The additions tend to be at deck level or higher, raising the center of gravity, so seismic boats tend to roll constantly.

Things that would, on other vessels, be solely the skipper's responsibility, are, on seismic boats, often controlled to fit seismic requirements. Navigation becomes a matter of staying precisely on a line, speed a matter of recycling time of the source, and frequency of pops.

Surveying

Locations of shot points are obtained offshore mostly by some form of radio surveying, the most common forms being satellite surveying and Shoran.

The surveying, instead of being done ahead of the rest of the crew, with locations marked, as on land, is one of the processes that go on simultaneously, offshore. The survey system guides the boat, with instructions, either verbal or on a screen, to the man at the wheel, to make corrections one way or the other, on the course of a line, and similar instructions to make an approach to the next line, so the cable will be out behind the boat when the line is started. Then, at frequent intervals, every pop, fourth pop, etc., the location of an antenna on the boat is recorded.

Precise locations from the readings recorded are not normally available immediately, but are produced later as a map by computer, at some center on land.

Pre-Plots

The offshore survey systems use pre-plots, shot point locations in a form usable by the particular system. They are usually calculated in advance by computer, for the entire program, and are then in the form usually of computer printouts. The printouts include coordinates of shot points and points leading to each line, to get started, for each end, as circumstances and planning in the field will determine which end of the line is best to start from.

If, during shooting, new lines are needed, there may be difficulty in getting pre-plots quickly. They would normally be worked up by computer, and delivered to the boat at the next opportunity. Or, a very limited amount of data could be transmitted to the boat by radio. This is very unsatisfactory, with interference and poor reception, and dependence on someone's writing a large number of numbers correctly. Also, for only a few lines, pre-plots have been calculated on board with a pocket calculator.

If there is a computer on board, the locations can be determined readily. Satellite surveying, one of the methods to be described, includes a computer.

Shoran

Shoran is a radio surveying system using two "base stations" at known locations to determine the location of a "mobile station". Each station has a transmitter and receiver. Each base station receives one signal from the mobile station, and immediately transmits it back. So the mobile station, by measuring the time from transmission of a signal to receiving it back, can determine the distance to a base station. With the two distances, its own location can be determined.

Shoran was originally a line of sight system, its name coming from short range navigation. But some of the radio waves are scattered by the troposphere, so part can be received beyond the horizon. The shoran used for seismic surveying utilizes this scatter, and is called extended range shoran, or XR shoran. It covers a range of about 150 to

200 miles in the tropics, diminishing to about half as far in the arctic.

Pre-plots for shoran are in "X-Y coordinates", distances in meters from arbitrary base lines chosen for the area.

To make the survey accurate, several techniques are used. The base station locations are determined as reliably and precisely as possible, by optical surveying, satellite, etc. The shoran equipment is calibrated by using it to measure the "base line", distance between base stations, and comparing that measurement with the known distance. During the program the boat crosses the base line when convenient, repeating the calibration. If more than two base stations are used, to cover a larger area, then at times a "three-way fix" can be obtained, to see if locations from different pairs of stations agree.

Shoran, like other triangulation, getting a location from two known points, works best when the three distances are about equal. Points forming an angle within 30° of the base line, are considered not very reliable.

Shoran can be recorded by extra equipment, or by people reading dials and making notations. The former is far more reliable. After a survey, the data are analyzed by computer and a location map prepared. This may take several weeks.

The equipment for the base stations is moved in to the locations, set up, an about 50-ft. antenna erected, a shelter set up, and an operator left there to keep the equipment running 24 hours a day for a month or so. He hires a local assistant or two, to haul things, etc. He may not know their language. So his social life is confined to his radio communication with the other base stations and the boat, maybe other similar networks. It isn't unheard of for an operator to have an additional, female, assistant.

Even so, it takes a special kind of person to be a base station operator, and enjoy it.

Satellite

Another form of radio surveying uses earth satellites. Several satellites are in polar orbit, so they pass over all parts of the earth. The system is Transit, Navy Navigation Satellite System, or NNSS. It is a U.S. Navy system, available to all users.

A satellite broadcasts two very stable frequencies, and includes in the tones data on the location of the orbit and a time signal. The orbit and time data is renewed by a station on earth about every 12 hours.

The boat has a receiver and computer to obtain locations from the broadcasts. The tones are received with a higher pitch as the satellite aproaches the boat, lower as it leaves, the doppler effect. And the change is varied more for a pass near the boat, is smoother for a distant pass. So the computer, from the frequency changes, orbit location, timing, can determine the boat's position.

Satellite passes can be as much as several hours apart, so an "integrated satellite system", adds some other types of information for dead reckoning by the computer between passes.

A gyrocompass, using the earth's rotation to give it a north-south alignment, provides the boat's exact heading.

Doppler sonar sends four pulses of sound to the sea bed, and receives them back, reflected from the sea bottom, or, in water deeper than about 600 to 2,000 ft., scattered back from the mass of water. Doppler effect makes the received frequencies different from those sent out, indicating the boat's speed. The sound reflected from the sea bed gives a much more accurate measurement than the scatter, as the scatter is from water that has current movements.

A velocimeter measures the velocity of sound in the water, to improve the doppler sonar data.

An inclinometer measures pitch, roll, yaw of the boat for additional corrections to the doppler sonar.

In deep water, Loran-C or inertial navigation can substitute for doppler sonar. Loran-C (Loran, *long range navigation*) is a system using base stations that covers a range of about 1,500 miles, but not as accurately as a satellite. Its signals though, can be used in a different way to give good data on, not location, but motion, of the boat.

Inertial navigation, with an inertial platform, uses two pendulums mounted at right angles to each other. Any change of speed of the boat in a forward, or sideward, direction causes the pendulum mounted in that direction to swing. The pendulums are on a gimbal-mounted platform that is held level by three fast-spinning gyroscopes mounted at right angles to each other. The gyroscopes' resistance to tilt in each of the three dimensions keeps the platform level, so their swings are not affected by the roll and pitch of the boat.

From all its data, the computer calculates locations of the boat between satellite passes. Then, at the next pass, gets an "update", a new satellite-derived location.

The computer, given the first and last points of a line, can guide the boat to an approach to the line, along the line, and signal where shot points are to be. The information is displayed in the wheelhouse, so navigation is just a matter of keeping a dial on zero.

Pre-plots then, need only be locations of end points of each line, in latitude and longitude, the coordinate grid used by the satellite system.

The data is recorded on tape. Later, in a computer center, better locations can be made, smoothing jumps at updates, etc., and a map made.

It may sometimes be necessary to stop shooting to wait for a good satellite pass, one neither too straight overhead nor too low on the horizon. Except for this problem though, the main advantage of satellite surveying is convenience. There are no base stations. The computer adds versatility, as pre-plots for new or changed lines can be determined by the computer and the lines shot immediately.

Another use of satellite surveying is to set up a portable satellite unit at a fixed location to be determined—well location, shoran base station, etc. A number of passes are combined to give a good location.

Other Survey Systems

Loran-C, as stated, is a system using base stations, that gives locations as far as 1,500 miles from the stations, but without the accuracy of satellite. It uses pulses of radio energy, or RF (radio frequency). Other systems derived from it are more accurate, but at the sacrifice of range. Some of them have been set up as a substitute for shoran, with higher antennas (around 300 ft.), permanent base stations, and about the same range as shoran. They are highly accurate, and the permanent setup allows the same system to be used for successive seismic programs, and also for well locations that may be made based on the seismic data, and which therefore need to be compatible with that data.

Phase-comparison survey systems use two base stations and a mobile station, like shoran, but instead of using measurements of time for the radio wave to go from one station to another, they depend on the interference between frequencies broadcast from the two shore stations. The interference pattern is such that different parts of it can be distinguished by the instrumentation of the mobile station.

The pattern is repeating in cycles. It is the different parts of a cycle that can be distinguished. Two different cycles cannot be distinguished from each other. A cycle forms a "lane", some width along a hyperbola. To avoid confusion, a "lane count" is kept from the last known point, just a count of how many lanes have been crossed. Or an additional set of radio waves with a different interference pattern, and less accuracy but wider lanes helps distinguish the primary lanes.

Hydrophone

The type of geophone used offshore is usually called a hydrophone. This is something of a misnomer, as it, like any geophone, obtains data about the earth (geo-), not the water (hydro-). There is another use for the word hydrophone, a device to listen to sounds in the water, fish, submarines, etc. However, marine geophones are quite different from those used on land, so a special word is necessary to make that distinction.

A hydrophone, pressure geophone, pressure phone is made to detect, not motion as in a land geophone, but changes of pressure. Most hydrophones use a piezo-electric material, that is, one that generates electricity when squeezed or released (piezo, from a word meaning to press). The compressions and rarefactions that are sound waves alternately squeeze and release, generating current proportional to the pressure changes. Pressure phones work best at a pressure existing some distance down in the water. They are built into the seismic cable (Fig. 3-20). and towed at a controlled depth of around 40 feet.

The type of pressure phone used on offshore crews is the "acceleration-canceling hydrophone", which has two piezo-electric units in it, so placed that any effects of motion on one unit, will be in the opposite direction on the other, so they will tend to cancel. This pretty well eliminates the effects of boat engine noise, cable jerks, etc.

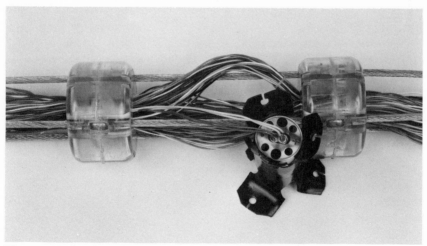

Fig. 3-20 Hydrophone in cable   Credit: GeoSpace Corp.

Marine Cable

A marine seismic cable, or streamer, is trailed behind the boat. The geophones, or hydrophones, are built into it, so all the connecting and spacing of geophones in groups is already done.

On long trips, and when going into port, the cable is carried on a large reel on the back deck. Then, at the start of a shooting program, it is payed out and is towed behind the boat throughout the shooting, day and night.

The cable is thicker than a land cable, with a protective clear plastic tube about 3 in. in diameter. In the tube can be seen the wires, colorful with their color coded insulation, loose, not wound tightly together. A steel cable in the center takes the strain of pulling the long, heavy cable through the water. The cable should be of about neutral buoyancy, that is, about the weight of an equal volume of water. Adjusting the buoyancy is called balancing the cable. It is done by filling the plastic tube with a kerosene-like liquid through valves in the cable, and by adding lead strips, taped against the cable, where needed. The balancing may take a good part of a day.

Also, "depth controllers" are put on the cable at intervals, about four of them on the cable. A depth controller looks like a little torpedo, about two feet long, with two short winglike fins. It separates lengthwise into two halves, hinged, so it can be opened and clamped onto the cable. The fins can be angled, like the diving planes of a submarine, to keep the cable at a desired depth, maybe 40 ft., during shooting.

The cable is in sections, so a faulty one can be taken out and replaced. Some sections have geophones, and others, "dead" sections, do not. The two kinds may be alternated to arrange the spacing between groups.

At the end of the cable, a "tail buoy" is attached by a length of rope. The tail buoy is usually a catamaran raft with styrofoam floats. It is equipped to be detected from the boat, by sight, radar reflector, radio responder, or radio transmitter.

The cable is usually from a mile and a half to over two miles long, although it may be shorter for special situations. 2,400 m, about one and a half miles, is a fairly standard length. While shooting, the cable, down in the water, is not visible. Looking hard, the tail buoy may be seen as a sparkle on the horizon.

When changing from one line to another, the boat runs in a large curve to get the cable aligned straight behind it for the new line. A similar big loop is run if, in the course of the line, something causes the shooting to stop. Then, when the cable is again straight, some of the line is re-shot to overlap the earlier part.

The boat runs day and night, different shifts of people taking over, and the frequent pop or bang of the energy source interrupted only for line changes or repairs.

Green Dragon

There is a temptation to paint teeth and other ornamentation on the tail buoy, especially as, if all the electronics fails, a bright color may help to spot it. It is a mile or more behind the boat, with no visible connection between the two. When the boat turns to make a loop, the tail buoy may be seen, going by in the opposite direction. So it appears to go where it will, somehow under its own power. One offshore crew reported that an official advice to mariners once warned vessels of a green dragon sighted in the sea.

Air Gun

The seismic source most used offshore is the air gun. It is a chamber that is filled with compressed air, which is released suddenly. The pop of the released air is the seismic impulse.

In offshore seismic work, a number of air guns are trailed behind the boat, with hoses extending from them to air compressors on deck. There are about four large compressors and around four to a dozen air guns. Air is pumped into the guns. Then the air in all guns is released at once. This is referred to as a pop. The pops occur at about ten second intervals, for the length of a line, which may be a matter of several hours. Then, during the run to the start of the next line, the guns are hauled on deck and serviced. Often there are two complete sets of guns, so one set can always be in working order.

The guns are usually arranged in a "tuned air gun array." That is, the guns are of varying sizes, spaced varying distances apart, and with varying time delays, all to cause the expanding air from the different guns to blend, making in effect one large, sort of flat, detonation. The different sizes of air guns produce bubbles that expand and contract at different rates, so the combination of them all, allows the bubbles to partially cancel each other.

The guns are trailed a little way behind the boat, at a depth of around

30 feet. They may be attached one behind the other, or may be trailed separately.

The effect noticeable from the boat is a rather sharp shock, somewhat as though someone hit the hull with a padded hammer. A few seconds after the pop, the water sort of boils up, and looks white from the addition to it of a lot of tiny air bubbles. Then, 9 or 10 sec. after the pop, the next pop occurs.

Marine Gas Gun

A gas gun fires a mixture of propane and oxygen, or oxygen-enriched air. It is fired in a chamber.

In offshore operations, an array of gas guns is trailed behind or alongside the boat, much as air guns are handled. A tank of propane is carried on deck. A tank of oxygen may also be carried, or a nitrogen removal process may be used to increase the percentage of oxygen in air drawn in from outside.

In one gas gun system, the explosion takes place in a heavy rubber sleeve that expands with the explosion, then contracts, letting the gases escape through a tube to the air. This somewhat reduces the expand-contract-expand bubble effect of releasing them in the water.

Other Sources

Several other seismic energy sources are used offshore to a lesser extent than air guns and gas guns.

Steam, from a boiler on deck, is suddenly released from a chamber in the water. The particular advantage of this source is that the released steam quickly condenses into water, so there is no volume of gas remaining to form a second expansion of a bubble.

Very small charges of explosive are fired in the ,water, at about the depth and in the same rapid succession, as other sources. It is somewhat more forceful, so is used where deep information is difficult to obtain.

Occasionally, in special circumstances, like some shallow water operations, larger charges of explosives are suspended from expendable floats and fired within a few feet of the surface of the water. The gases blow out of the water before a second bubble expansion can take place. A geyser of water shoots up. This was the original source for offshore seismic exploration, and was the only one used for many years. Its use is now forbidden in many areas.

Recording

Offshore recording uses the same type of magnetic recording equipment as is used on land, but planned to fit the requirements of shooting in rapid sequence, on a 24 hour basis, and at sea.

There is more equipment. There may be two or more recording units, so the switch to a new reel of tape can be made without halting the operation while reels are changed. There is a larger stock of tape. There are indicators for cable depth, etc. There are enough personnel to work on a shift basis. There are spare parts, and room, to make repairs while shooting is going on, or on the run between lines.

So the "doghouse", recording room, is fairly large, although it may be crowded with equipment and people. It is air conditioned, not primarily for the comfort of the people, but so the heat-producing electronic equipment is kept at a good working temperature.

In normal operations, several people are in the doghouse, watching instruments, making notes, maybe repairing things or improvising to make things handier. There is a dull boom, not loud, of the air guns or other source going off. The tape reels advance a few inches as an oscilloscope shows the motion of the traces as it occurs. In less than 6 sec., usually, all the instrumental activity of that shot is over. People are still writing about it on logs of the operation. Then, about ten seconds after the boom, there is another boom. Another shot has gone off. The tape advances.

The logs give such details as depth of the guns, depth of the cable, time of day, sea state, shot number.

After a number of shots are recorded, the take-up reel is full. Someone switches to another recorder already loaded with a full reel and empty take-up reel. The full tape is taken off the machine, put in its case, labeled, sealed, put in a sack, placed in the carton the tape came in, notes made about that reel of tape. Someone puts another full reel of tape on the recorder, threads the end through a number of tension rollers and onto the empty reel, now put in the take-up position.

Meanwhile, more shots, on and on for hours, to the end of that line. Then a line change, with the boat swinging in a big loop to get the cable straight behind it for the start of the next line.

The recording people are also the ones who look after the cable, laying it out, balancing, testing. And, when an instrument in the doghouse indicates that the cable has a problem, they come boiling out to the back deck to reel it in to the bad spot, see what the trouble is, repair it, run the cable back out. To facilitate communications, there is usually an intercom connecting recording room, surveying, wheelhouse, and back deck.

In the recording room, when things are running smoothly, at shift-change time, maybe in the middle of the night, the new crew wander in early carrying cups of coffee and pieces of pie. They find out what has happened and how things are going, take their places. The shift going off stand around a while, glad to be off duty, but sometimes a little reluctant to leave the one place on board where the action is.

The Signal and Its Signature

The "signal" is the sound that starts out on its trip into the earth. It is distinguished from the response that comes back to the surface. The signal is made up of whatever sound the source produces. It can be measured by putting a geophone near the source and recording the sound as it leaves. This recording is called the "signature" of that source. The signature then, shows what energy was introduced into the ground, whereas the record section shows what came back to the surface. The signature is not the same as a wavelet, which is the same signal, but after going down into the earth and coming back up, filtered by the earth.

The main importance of seismic signatures is in offshore work, where the bubble effect in the water, the expansions and contractions of the gas released by the explosion, is a major problem. The signature shows the initial impulse and the bubbles following it, with their relative strengths, all being parts of the signal. The whole thing is put into the earth. The goal in source design is to get a strong initial impulse with weak bubbles, or better, no bubbles.

The signature of a source is measured by recording it with a geophone just a few feet away (near-field measurement) and also by a number of geophones more than a boat's length away (far-field). Deep water is required for these tests, so the reflection from the sea floor will not appear on the record before the bubble effect is recorded.

The appearance of a normal signature is a trace with one strong sharp peak and trough, followed by several others of decreasing strength, until they get too small to be noticed. This signature is the signal that is put into the earth, so any reflecting horizon bounces the whole thing back. A seismic record that is made up of these long duration reflections, has a lot of repetition in it, and overlapping reflections. The more the first energy of each reflection can be made to dominate, and the more the following parts can be subdued, the clearer the section will be. So it is desirable to design sources to have a strong first impulse and relatively weak bubbles (Fig. 3-21).

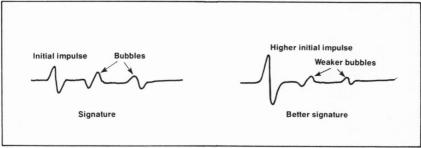

Fig. 3-21 Marine signatures

Telemetry

For special purposes, geophones are attached to telemetric equipment—radio transmitters to send the seismic data from the geophones to the recording equipment. The arrangement is applicable in situations in which the seismic cable is a hindrance.

A transmitting buoy may be set in water for each trace of a complete set of 24 or more traces. The geophones of the group are connected to the transmitter by wire (Fig. 3-22). Then a seismic boat can run, firing its source of energy, and receive data from midway between the geophones and source. Data can be obtained from under an island; or from very shallow water and onto the beach, to tie to lines shot on land.

Transmitters with geophone groups can be set out in jungles, reducing the amount of trail to be cut and making difficult access easier by eliminating the laying out continuously of cable. In some cases, it may be convenient to set out geophone groups with transmitters by helicopter.

Shallow Water, Surf, Mud Flats

The transition between land and deep water, the two main shooting areas, is a problem zone. Special field techniques are needed to cover this zone, and to link the land and water shooting together.

There are special shallow water crews, on shallow draft boats. If the water is too shallow for the normal marine sources to work well,

Fig. 3-22 Telemetry–a group of hydrophones being laid out from a Telseis*
transmitter   *Credit: Fairfield Industries, Inc.*

dynamite may be fired in the water, or holes may be drilled in the water bottom, and the dynamite fired in the holes. Surveying may be the usual radio surveying of offshore work, or may be by transit or theodolite from stations on shore.

An essentially land-type program may be carried out on mud flats, with people walking or wading, depending on the tide, drilling holes, placing charges in the holes, laying out cables and planting geophones.

The surf knocking against geophones is a source of noise, but can be pretty well eliminated by filtering out those frequencies.

Engineering Surveys

Engineering surveys, most often run for pipeline or rig locations, include several kinds of information about the sea bed and shallow subsurface. They are also sometimes run as an aid to interpreting a standard offshore seismic program.

A depth recorder uses reflected sound waves to show the depth of the water under the boat. Sometimes when there is a soft bottom, it may also show a harder layer underneath.

Side-scan sonar is a sound-reflection system that gives three-

---

*Telseis is a registered trademark of Fairfield Industries, Inc.

dimensional data. Sound from a pinger fans out, reflecting from a swath of the sea bed. A "fish" towed behind the boat receives signals from both sides. These are recorded on a continuous strip, showing, usually, all the sea bed from under the boat to 500 ft. to each side (Fig. 3-23). Rocks, ripple marks, sunken vessels appear more or less as on an air photograph, but "illuminated" from the center line. Shadows are white, and extend outward from the objects casting them. Height of an object can be estimated from the length of its shadow. Rulings are made along the record to indicate distance to the side of the boat. They are in 50 foot increments. The record can, at a different setting, be made to cover only 250 ft. of sea bed on each side, showing more detail; or 1000 ft. to cover more area with less detail. In each case the lines are placed so they represent 50-ft. widths of sea bed. Some equipment is metric, with 25-meter increments and a 500-meter width.

Side-scan sonar tells much about the bottom; hard or soft; mud, sand, or rock; etc. If run simultaneously with a seismic survey and if the bottom has recognizable features, it can be a check of surveying on line

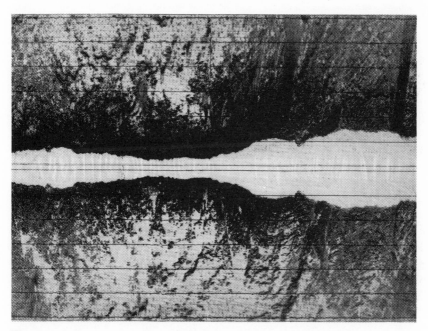

Fig. 3-23 Side scan sonar record off U.S. West Coast showing volcanics
Credit: E.G. & G. Environmental Consultants

intersections. It introduces to offshore work the "landmarks" that are so taken for granted on land.

Side-scan sonar can show clearly the bubbles of gas seeps. This helps in locating near-surface hydrocarbons and serves as a warning against the hazards of drilling into shallow gas.

Side-scan sonar has an extra advantage when run simultaneously with a seismic survey, to show the configuration of the bottom along the seismic lines. Each time there is a pop of the seismic energy source, it makes a strong line across the sonar record. This interferes a bit with the picture of the sea bed, but the interference is more than made up for, by the definite tie between points on the sonar record and the seismic pops. Every so often, a pop must be identified by the operator's marking its number on the sonar record. Then, for any pop, there is a map of the surface, somewhat like an air photo onshore. If the sea bed is not just mud or sand, but has some recognizable features, the sonar tape can help a great deal in identifying locations, checking intersections of seismic lines, positioning wells, etc.

The electrical spark method uses sound of higher frequency than normal seismic energy to yield very detailed information for a relatively shallow depth below the sea bed. In general, the less penetration the more detail. The very shallow surveys can show remarkable detail on small faults that extend up to the sea bed. This is particularly useful in areas where shallow gas pockets can be a hazard to drilling.

All these shallow subsurface methods use very short geophone cables—necessarily, as the geometry of shallow data does not allow for long offsets. That is, a long cable, or rather, the more distant parts of a long cable, will not receive reflected energy from shallow horizons. The short cables are generally referred to as mini-cables. Often such a near-surface survey is run at the same time as a normal seismic survey. The mini-cable may be, for its entire length, nearer the boat than the nearest geophone group of the regular cable. The mini-cable and its associated equipment record data about the near-surface in fine detail. The normal seismic cable obtains the usual data, with maybe faults, or structure, approaching the surface, but not clearly indicated at the top of the section. The mini-cable data then can provide interesting answers to geological questions. A fault may be seen at the surface, and be deduced to be the same fault as one on the normal section—projecting downward on the one section and upward on the other. Or, equally useful, the deeper fault may be seen to not reach the surface. Similarly, structure may or may not be found to reach the surface.

Gravity and Magnetometer

Gravity and magnetic measurements are not, on land, a part of seismic exploration, but in general are preliminary, to isolate prospects for seismic investigation. But, offshore, to take advantage of the transport and the surveying and other measurements that are being provided anyway, one or both may be an integral part of the seismic operation. In this application, they are supplemental to the seismic work. Different approaches, not having the same defects, can serve as a check on each other.

A gravity meter, or gravimeter, measures the weight of an object in the instrument. If dense rocks are nearer the surface than normal in one location, the object in the instrument will weigh more there than at another location. So the gravity meter indicates places where denser, usually older, rocks are near the surface, therefore some indication of structure. A structural high appears as a high gravity reading, a "gravity maximum." The low density of a salt dome shows up as a lesser gravity value, a "gravity minimum."

On a boat, a gravity meter is placed on a surface that is level regardless of the motion of the vessel, like the level platform of inertial navigation. The rise and fall of the boat and other effects of its travel are measured by various instruments, to later be taken into account in computer processing.

A magnetometer is, in effect, but not in construction, a magnetic compass on its side. That is, it measures variations in downward magnetic attraction at different locations. Igneous rocks are more strongly magnetic than sedimentary rocks, so a magnetometer gives information on depth to basement, the bottom of the sedimentary section.

On a boat, the magnetometer is trailed in the water. Data is transmitted through a cable to a recording device. The recording may be on paper or magnetic tape.

4

Digital Recording

Analog and Digital

Seismic data, or any other kind of data, can be recorded and processed by analog methods, or digitally.

Analog refers to representing an amount of something by an amount of something else. A thermometer, a clock hand, a seismic trace on a record section, are all analog representations. Temperature is represented by the height of the liquid column, time by the position of the hand, movement of the earth by the swings of the trace.

Having a quantity in digital form just means it is in numbers, digits. A weather report gives temperature digitally. A calendar is digital. A seismic trace is represented digitally by a long series of numbers.

Seismic exploration was completely analog when it began. The ground shakes, a voltage is generated by a geophone, and a trace wiggles. When magnetic tape recording was developed, it enabled the trace to be replayed as though the explosion had occurred again, and the trace was re-recorded onto a record section. Still analog. The trace could be modified with different electrical processes, filtering out unwanted frequencies, changing amplitudes, shifting or stretching the trace.

Digital seismic recording seems at first like a backward step, in that it leaves gaps in the data. At intervals, typically every two or four milliseconds, the amplitude of the trace is recorded as a number. The parts of the trace in the gaps between the samples are lost, whereas in analog recording, the trace is continuous. However, the trace can usually be considered to be a smooth curve between samples. In exchange for the one loss, there are things that digital recording provides that analog does not. Digital recording has a much greater dynamic range, range between the largest and smallest amplitudes it can record, than analog recording. And with the data in the form of numbers, modifications can be performed by arithmetic, so a trace can be replaced by another trace that is the result of arithmetic operations performed on the original trace. The things that can be done arithmetically only require a formula to be worked out and put in the form of a computer program, so a computer can do all the arithmetic on all the traces, its high speed making the work feasible.

Aliasing is a characteristic of digital recording, as it is of any method of sampling data at intervals. It is the false picture you get if you don't sample often enough. And the finer the detail, the oftener the need to sample to avoid an alias. If seismic data is sampled less than twice per cycle (peak and trough combination), that is, if there isn't at least one sample on each peak and one in each trough, then, when the wave form is later reconstituted from the sample, it will have a different, lower, frequency (Fig. 4-1).

So, for a certain sample rate, say 4 milliseconds, the higher frequencies will be aliased. To prevent this problem, an "alias filter" is applied to the data, to eliminate those high frequencies before the data is sampled and recorded.

Binary Numbers

Digital recording of seismic data uses a binary number system. That is, a system to the base two. Our usual arabic numerals (used by nearly everybody in the world except the Arabs—but derived from theirs) are to the base ten. We have a different digit for each number smaller than

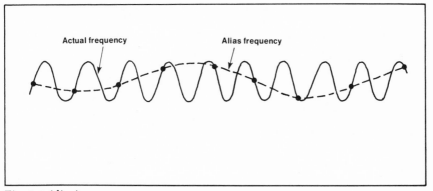

Fig. 4-1 Aliasing

ten, and to make the number ten, use the digit for one, followed by the zero digit. Then for higher numbers, go on through 11, 12, 13, etc. Presumably, the base ten was chosen because we have ten fingers. If there was a tribe somewhere that didn't use their thumbs in counting but used a similar system, their numbers might be 0, 1, 2, 3, 4, 5, 6, 7, 10—the 10 (don't call it "ten", but "one-zero", or "eight", or something) representing the number of fingers on two hands, the number we call eight. "20" would be the way to write sixteen, "25" would be twenty-one, etc.

Following this idea, a number system to any base can be devised. All you need is enough different digits for all the numbers up to the base, then use 10 for the base, 100 for the square of the base, etc.

Numbers to the base ten are called decimal. Other types are useful. Duo-decimal, base twelve, can be divided up into more whole-number parts, so we buy eggs by the dozen. We can get a half, third, quarter, sixth of a dozen without breaking an egg.

Octal, the base eight already described, is handy for some computer uses.

Binary (we can't call it dual, there's already a meaning for that word), base two, is an especially useful system. Like the other systems, it has digits for all the numbers below the base. That is, it has the digits 0 and 1. So the numbers go like this.

Binary	Decimal
0	0
1	1
10	2
11	3
100	4
101	5
110	6
111	7
1000	8
1001	9
1010	10
1011	11

Now this binary system uses a lot more digits to write a large number than the decimal system does, but it has the big advantage that it needs only two different digits, 0 and 1. This makes it admirably suited to digital computer operations. The 1 can be represented on magnetic tape by a magnetized spot, and the 0 by a spot that is not magnetized. Then with the simple convention of dividing a tape into specific spots, and either magnetizing them or not, any number can be "written" on the tape. This is much more definite than a system, like analog recording, that uses degree of magnetization. With digital binary recording, a spot is either magnetized or it isn't.

Some special number systems, also using the binary on-or-off principle, are used in computers. One such is binary coded decimal, which writes in decimal form, but by expressing each digit of the decimal number in binary form. Four spaces for each decimal digit allow it to be written in binary.

Exponential Notation

In dealing with very large or very small numbers, a lot of zeros get in the act. 100,000,000 or .00000001 have so many zeros that you have to count them to decide what the number is. There is a different system of notation that includes a number to indicate how many zeros there are, like "1 followed by 8 zeros", or how many decimals, like "1 in the eighth decimal place". Also, the system is so designed that it takes care of more complicated numbers, like 132,000,000. It is the exponential system in which that last number is 1.32×10^8.

The exponent, the 8 in that example, is the power of the number ten, the number of times ten is to be multiplied by itself. 10^2 is 10 x 10, or 100. So a million is 10^6, and a hundred million is 10^8.

Negative exponents represent dividing by 10. 10^{-1} is 1/10 or .1. And 10^{-2} is 1/(10 × 10) or .01. And 10^{-8} is 1/100,000,000 or .00000001 or "1 in the eighth decimal place". The negative exponent is the number of zeros in the denominator of the fraction 1/1,000 or one more than the number of zeros in the number when expressed as a decimal (.001).

A tabulation will help show the relationship.

Exponential	Decimal		Exponential	Decimal	
1×10^3	1000		4.2×10^3	4200	
1×10^2	100		4.2×10^2	420	
1×10^1	10		4.2×10^1	42	
1×10^0	1		4.2×10^0	4.2	
1×10^{-1}	.1	(1/10)	4.2×10^{-1}	.42	(4.2/10)
1×10^{-2}	.01	(1/100)	4.2×10^{-2}	.042	(4.2/100)
1×10^{-3}	.001	(1/1000)	4.2×10^{-3}	.0042	(4.2/1000)

4.2×10^{16} 42,000,000,000,000,000
                                  (16 digits after the 4)

4.2×10^{-16} .00000000000000042
                                  (decimal and 15 zeros)

The exponential system is convenient for writing very large and very small numbers. It doesn't take up as much room on a page, and it is easier to read and comprehend.

It is handy also for labeling charts, especially ones that have so much range of large to small numbers that they are plotted logarithmically. It is used in pocket calculators to allow a limited display area to handle a greater range of numbers.

It is also referred to as "floating point", and allows seismic computers to handle a greater range of numbers.

Binary Gain and Floating Point

The amplitudes of seismic energy go through a wide variation, from very large at the initiation of the energy to extremely small a few seconds later. When dynamite, air gun, etc., is fired, there is a loud noise, suddenly, of the explosion. That is all that a person can hear, but for seismic use we need the reflected sounds for several seconds after that. Also a record section would not be clear to human eyes if the amplitudes on it were in their true relationships. The strong energy at first would overlap from one trace to another into a hopeless tangle, and the deeper reflections would mostly be indistinguishably weak wiggles on dead-looking traces.

So the early energy needs to be cut down to size, and the later amplified to easy visibility. This is gain control. A very early type was the observer's hand on a knob. He started with the gain set very low, and after the shot, for the next two or three seconds, steadily, he hoped, advanced the gain. That is like the early movies, shot with a hand-cranked camera. The cameraman had to avoid getting excited about the action, or he'd crank faster, slowing the action when it appeared on the screen.

This gain control was replaced by other methods—a steady increase by machine, called a programmed gain control; a device that cut the gain down right after any burst of energy, and let it come back up till the next burst. This latter was AGC or AVC—Automatic Gain Control or Automatic Volume Control.

All this fiddling with the gain helped to make more usable traces, that showed both strong and weak energy pretty well. The records could show both shallow and deep reflections. But these gain controls, in making the record more uniform, destroyed the information on relative strengths of the reflections.

Binary gain is a type of gain control applied in recording digitally. It is called binary, because digital computers use the binary number system. That is, each switching device in the computer has only two positions, on and off, so all numbers must be represented by the combinations of the digits 0, or off, and 1, or on. A trace is recorded, starting with a certain gain. Then, at frequent intervals, the amount of energy being recorded is measured and the gain raised or lowered, as needed. A zone of either strong or weak energy is recorded at an appropriate amplitude, and a record is kept of the amount of gain used to bring it to that level. This record of the gain used preserves a knowledge of the absolute strength of the original energy reflected. From this information, relative strengths of a reflection on successive traces can be compared. This preservation of relative amplitude makes possible the bright spot technique, which indicates gas in formations.

A gain system more advanced than binary gain is floating point recording. It uses binary numbers, but in exponential form, so it can handle a much greater range of numbers with the number of bits available on the magnetic tape. Very large numbers are written with large exponents, very small numbers with large negative exponents. So the recording can include all the seismic amplitudes, from the strongest to the weakest, that are at present usable. No gain changes are required, and relative amplitudes are preserved directly in the data.

Recording Format

Most seismic data is recorded on half-inch magnetic tape. The tape is recorded in 9-track format, that is, there are nine bits of information in the width of the tape. The tape recorder has 9 recording heads in a half inch of width. The tape is on reels, 8 or 10½ inches across, containing 1,200 or 2,400 ft. of tape. The reels are carefully protected from dirt. They are in individual holders, usually of clear plastic, which fasten closed with a turn of an inset handle. The holders in turn are contained in sealed plastic bags. The bag is cut open, the holder opened, and the reel put on the recording equipment and the tape threaded. With each shot, a few inches of tape are recorded, and the tape advanced that far onto a takeup reel. When the entire tape is recorded, the full takeup reel is taken from the recorder, put in the holder, which is closed, labeled thoroughly, and, on some crews, even put into another plastic bag, which is heat-sealed closed.

An older system, 7-track tape, also half inch, is used by some equipment.

One-inch tape is also in fairly common use. It is recorded with 21 tracks of information in the 1 inch of width.

At the time of shooting, field notes are kept on paper forms. They relate shot point numbers to tape reel numbers, etc., and have notations of circumstances in the field, weather, sea state, truck problems, etc.

Multiplex

Early, analog, magnetic tapes were wide, maybe four inches, and were about two feet long. One tape contained the data from one shot, and had as many channels as there were geophone groups, usually twenty-four. The tapes were therefore very like paper records, but in magnetized form, on tape.

Present day digital recording tapes are a half inch or one inch wide, and have seven, nine, or twenty-one channels. So how can they accommodate the recording of twenty-four, forty-eight, or more traces? They do it by taking samples of the traces in turn, or multiplexing. The data is sampled at intervals of one, two, or four milliseconds. So a sample of the first trace is recorded, then a sample of trace 2, etc. Each sample requires a number of bits to make it up, so that sample may run through all the available channels, (leaving some for other information) maybe one or two and a fraction times. After a sample of each of the traces has been recorded, a millisecond or more has elapsed, and another sample of trace 1 is recorded, etc.

The field tapes are this sort of mixture of traces, which will later be sorted out, demultiplexed, in the processing center.

5

Data Processing

In seismic work, improvement of the usefulness and quality of data is achieved by a number of processes, some fairly simple, others more specialized or sophisticated.

Corrections must be made for topographic variations, weathering differences. Filtering to retain only the best frequencies can be used. Normal moveout can be removed. Reflections can be more sharply defined by deconvolution. The records can be put together into record sections.

Demultiplex

The first step in processing the data is to get the traces separated from each other, so they can be worked on. In the field they had to be recorded all together as the first samples of all the traces had to be recorded in the time of 1 to 4 milliseconds, so the equipment would be ready to record the second sample. But the processing needs a single trace, separate from others, to work on.

So the data is demultiplexed. The first sample of trace 1 is located on the field tape and recorded on another tape. Then the second sample of trace 1, etc., until the entire several seconds of trace one are recorded on the new tape. If a record is 6 sec long, and sampled every millisecond, there are 6,000 samples. Sampling every two ms produces 3,000 samples, every 4 ms, 1,500 samples. Then trace 2 is put on after the last sample of trace 1.

The words multiplex and demultiplex, in processor's shorthand, have been reduced to mux and demux.

Near-Surface Corrections

In seismic work on land, there is a topographic problem. If the ground surface is absolutely flat, then seismic lines can be plotted below a horizontal line representing the surface. But if there is a hill, reflections so plotted will appear to have a lowered, synclinal shape under the hill (Fig. 5-1). A valley on the surface will produce an apparent elevation, or anticline, on reflected horizons. The surface feature affects the record all the way down equally, as the surface is the same no matter how deep a reflection is. So the entire traces need to be moved up or down to take out this effect.

How much should the traces be moved? It sounds simple at first. If a hill is 40 ft high, move the traces up 40 feet, so their upper ends follow the curve of the hill. But the scale of seismic traces is in time, so there needs to be some converting done to find out how much time represents 40 feet. This is usually done in several parts, and includes the problems of weathering and the difference between the elevation of a shot in a shot hole and an instrument on the ground.

If the shot is fired in a hole, the sound starts some distance below ground, but is received on the surface (Fig. 5-2). To even up this lopsided situation, the time spent at the receiver end of the travel path is reduced to bring it to the level of the shot.

The time missed by the depth of the shot is measured by a geophone, the uphole jug, shot point seismometer, SPS, placed near—within about ten feet of—the shot hole. The sound from the explosion is recorded by this uphole jug and the time from shot to surface measured. The instant of the shot is displayed on the record as a sharp break on a trace. The uphole jug makes another break on a trace when the sound arrives at the surface. The time between these two is measured on the record. This is the "uphole time", t_{uh} (Fig. 5-3). It is the time the sound took to go through the material alongside the hole. This uphole time can be assumed to be the same as the time to the same depth at the receiver, so, subtracting the uphole time from the total reflection time, cuts off that extra time at the receiver end (Fig. 5-4).

Shot holes are intended to be shot a little below the base of the weathered layer. If this is done, then time spent in the weathering is also removed by subtracting the uphole time.

If the shot is in the weathering, or if it is on the surface—i.e., if no hole is drilled—corrections for the weathering or additional weathering can be made, based on first breaks.

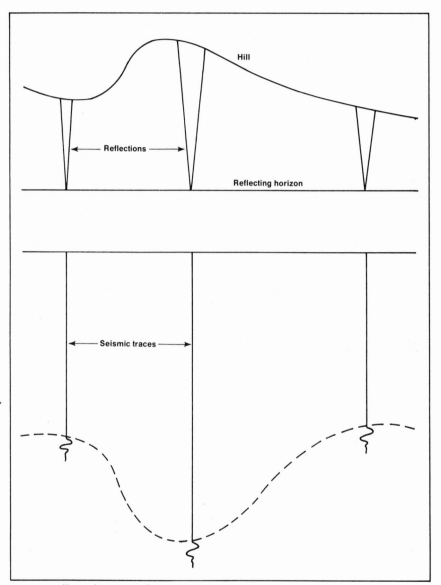

Fig. 5-1 *Effect of topography*

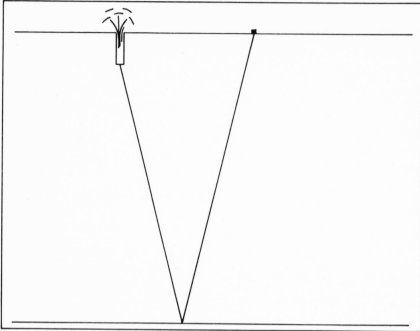

Fig. 5-2 Unequal paths

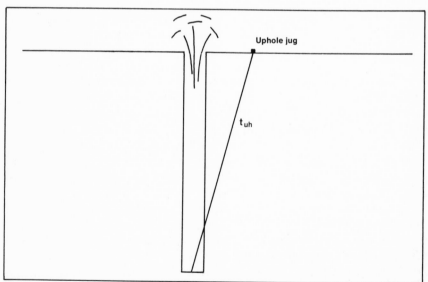

Fig. 5-3 Uphole time

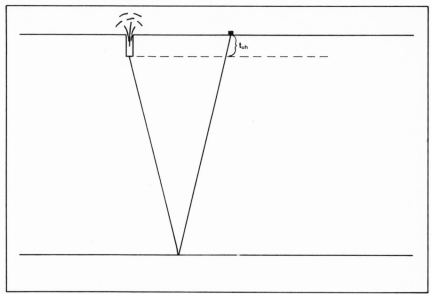

Fig. 5-4 Extra time cut off

First breaks are the earliest indications of energy from the shot on the various traces. Their times after the shot are results of the distance of the instruments from the shot point, and the velocity of sound through the route it takes to the geophones. The sound goes to a fast layer, is refracted, bent, along it, and refracted back up to the surface (Fig. 5-5a & b). This route tends to be quicker than a direct path through the weathered material, as the weathering, with its low velocity, transmits sound slowly. So the faster route, even though longer, gets the first energy there. Plotting these first break times on grid paper or treating them similarly by a computer program, allows the time to the high-speed layer to be determined. So the amount of weathering can be corrected for, by subtracting time spent in the weathering (Fig. 5-6). The correction is covered more thoroughly in Chapter 9, Refraction Exploration.

The datum correction is simple. A velocity is assumed for the vicinity of the datum plane, and used to correct the time to the datum. If the datum is below the depth already corrected to, then the time on down to the datum and back is subtracted (Fig. 5-7). If the datum is shallower, the time is added. The reflection time remaining is time spent below the datum, so variations in this reflection time can gener-

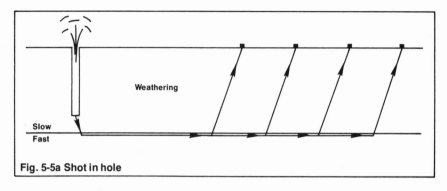

Fig. 5-5a Shot in hole

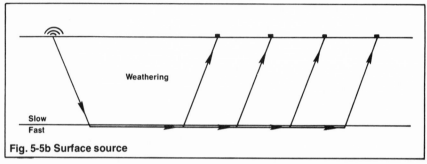

Fig. 5-5b Surface source

Fig. 5-5 *Weathering refraction paths*

ally be ascribed to variations in the horizons. The datum elevation is chosen near the elevation of most shots, so the use of the assumed velocity is kept to a minimum. These corrections are made for each shot point.

These corrections are, on land crews, often made by hand in the field, where questions may be asked, surveying checked, etc. The computations are sent to the processors along with the field tapes.

Static Corrections

The near-surface corrections calculated in the field are, in the processing center, put in the form of instructions to the computer to move each trace up or down the number of milliseconds called for by the elevation and weathering calculations.

These statics serve to smooth out the reflections to the extent that the hills and valleys are no longer discernible in reverse on the reflections,

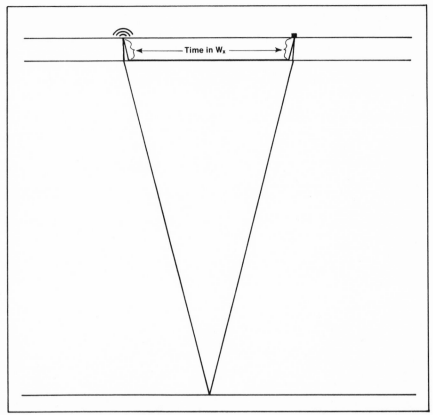

Fig. 5-6 *Weathering correction*

and that the weathering is fairly well corrected. However, they leave some irregularities that prevent the reflections from lining up cleanly, and combining well for CDP.

Additional correction can be obtained by automatic statics. They are calculations made by computer, using the large amount of data available from common depth point shooting. There are a lot of energy paths from each shot, and a lot to each geophone group, so statistical analyses can be made for these points. They can be arranged in different ways, and made to yield relative static fine corrections, which can be combined with the corrections already made.

The automatic statics use reflections to make the corrections, so they depend on there being good enough reflections to be lined up. Therefore, a section in an area of good reflections, but irregular near surface

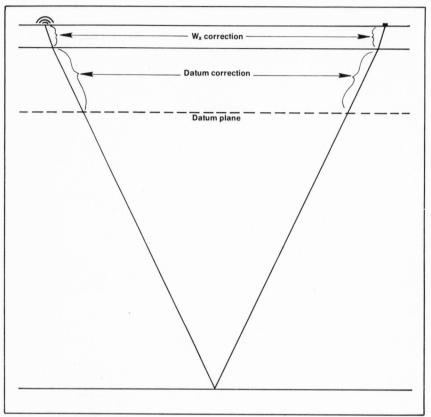

Fig. 5-7 Datum correction

situation, can best be improved by automatic statics. If reflections are poor, the method won't work well. If the near surface is smooth, the traces will already line up well.

The type of improvement to be seen in a good example of what automatic statics can do, is in the appearance of reflections on the final section (Fig. 5-8). A section with ragged looking, difficult to pick, reflections, can, in an optimum situation, be corrected so the reflections are smooth and clear.

100% Section

A 100% section may be made during data processing, to show how some of the steps are working. The section is made by displaying all the

Data Processing

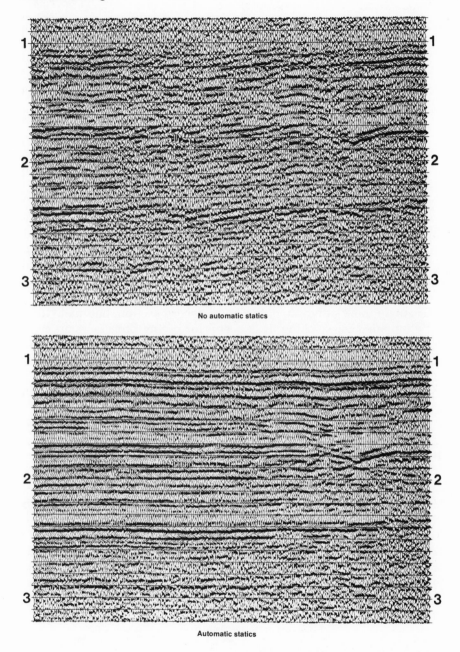

No automatic statics

Automatic statics

Fig. 5-8 Automatic statics   Credit: Western Geophysical Co.

traces of only certain records, enough to cover the length of the line only once, with no multiplicity for stack. Thus, if the data is 2,400%, then every 24th shot point is shown. A characteristic of this type of section is that the first breaks show, slanting down from each of the shots displayed, giving the top of the section a saw-toothed edge. This type of display is useful mostly to the processors.

Single Trace Gather

A gather is a selection of traces from among all those recorded on the magnetic tape, usually thought of in the form of a visual display.

A single trace gather is a display of a seismic section made up of just one trace from each shot. It may be a near trace gather, using the trace from the group nearest the shot, or a far trace gather, using the most distant one, or may use any trace in between, but usually the same trace from each shot, so normal moveout will not cause differences in reflections from trace to trace. These gathers are usually made before normal moveout has been removed.

A single trace gather then, is a 100% record section, but made from data intended for stacking. It is useful for giving processors an early look at the data, to help in the processing. A look at it may show which are the stronger reflections, whether there are steep dips, something about multiples.

A special purpose single trace gather is made on the field crew, to show what kind of data they are recording. It is a far cry from a processed section, and doesn't have very good data for interpreting, unless the area has very good record quality, but is a useful monitor of the shooting.

A single trace gather in processing may be used as a check, of one type, on the processing. Stacking of data is solely for the purpose of making the data better than in a 100% section. So if a gather has a valid reflection showing up better than the stacked section, there is a problem. Multiples, though, should look poorer on the stacked section.

Some interpreters like to have a single trace gather of each line, to help in interpreting. However, this is somewhat expensive if it must be ordered as an extra display. If it is a normal byproduct of the processing, then there is no extra expense. Different processing centers operate in different ways, so it is a byproduct in some, and not in others.

CDP Gather

Another useful type of gather is the CDP gather. It is a display of all the traces for a common depth point, all the traces that will later be combined to make a single stacked trace.

It looks very like the record of a single shot, but with some important differences.

A record of a shot has its traces side by side. The trace nearest the shot may be at the center of the group or at one edge, depending on the kind of cable and shot layout used. The first breaks slant down from the traces nearest the shot. The reflections cross the record with normal moveout curvature, and with an angle caused by the dip of the reflecting horizons.

The CDP gather is a similar display of traces, but selected as the traces with a common depth point. It has first breaks that slant from the near traces, and reflections that curve with normal moveout. But the reflections do not slant with dip.

So the CDP gather has normal moveout, but not dip, and therefore can be used to determine the amount of normal moveout curvature in the record. This is done by applying a correction to remove the NMO, and then inspecting a CDP gather to see how well the NMO has been removed. If it has been exactly removed, the reflections, with neither dip nor NMO, should line up exactly straight across the gather. This has application in determining velocity and in stacking data.

Stacking

Stacking is the combining of two or more traces into one. This combination takes place in several ways, and for several purposes.

The simplest way of combining traces is just to take the wires from two geophones, and twist them together, so just one set of data is recorded. This kind of blending can take place near the geophones, or at any point in the recording or processing. It can be after the energy is amplified, or after the traces are corrected for elevation differences, etc. The word "stack" is usually reserved for the combining during data processing.

In digital processing, the amplitudes of the traces are expressed as numbers. So stacking is accomplished by adding the numbers together.

In either way, two peaks on two traces will combine to make a peak, as high as both added together—if the two peaks are at the same times on the traces (Fig. 5-9). If they are at different times, the combined trace will have two separate peaks the size of the original ones (Fig. 5-10). A peak and trough lined up will tend to cancel each other (Fig. 5-11). If their amplitudes are equal, the combined trace will have no energy at that time.

After combining, the new traces are "normalized"—that is, the amplitude is reduced so the doubly high peaks created by reinforcement will be of normal height. Then a peak that was on only one of the traces, not having been reinforced, will be reduced by the normalization (Fig. 5-12).

Traces can be stacked for different reasons. Stacking can be a test of NMO corrections, to determine velocities in the subsurface. It can be used to just combine two adjacent traces, so they can be treated as one trace in processing, to reduce the amount of processing required for a large number of traces. The most common use of stacking is the combining of traces in common depth point processing. This is so well known that "stack" tends to mean "CDP stack". So "six-fold CDP" and "600% stack" are synonymous.

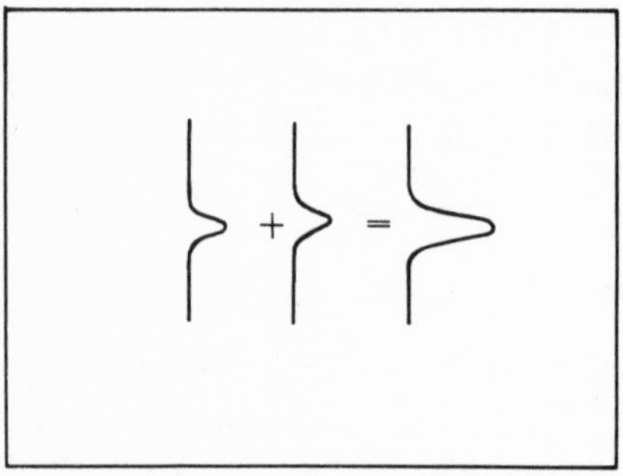

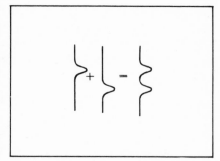

Fig. 5-9 *Two peaks at same reflection time*

Fig. 5-10 *Two peaks at different times*

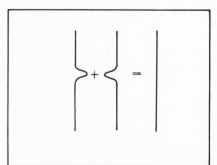

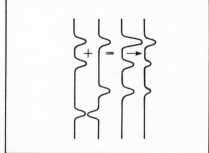

Fig. 5-11 *Peak and trough combined*

Fig. 5-12 *Normalized trace*

Multiple Attenuation

There are some slight differences in the geometry of primary (one-bounce) and multiple reflections that make it possible to weaken the multiples somewhat.

To see how multiples in general are attenuated—not eliminated—suppose a primary and a multiple occur at about the same place on a record section. The primary energy goes down to the horizon and back up to the surface. The multiple energy goes to a horizon about half that deep, up to the surface, down and up again. So the primary traverses a distance twice—down and up, but the multiple covers only half that distance, but four times (Fig. 5-13). But the two used about the same time to travel, so occur near each other on the record section.

In general, the velocity of sound increases with depth. Sound goes faster in the deeper rocks. So the two-way trip of the primary and the

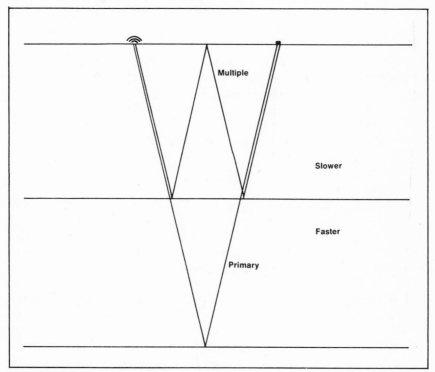

Fig. 5-13 Primary and multiple at same time

four-way trip of the multiple don't use the same velocities all the way. The lower half of the primary's course takes it through a faster zone than the multiple ever gets down to.

Now the normal moveout phenomenon on a record from the field—the curving of reflections because of the greater distance to the farther geophones—is dependent in part on velocity. The slower the velocity the more moveout. So a multiple which stays in the slower rocks, will have more of this moveout, or curving, than a primary at the same time on the record.

This difference is used to discriminate against multiples on record sections. When corrections are made to remove normal moveout, velocities are part of the corrections. That is, the moveout for a certain time at a certain velocity is calculated, and the traces corrected that amount to make reflections line up straight across, with no moveout. But if the velocity is wrong, the correction will not be right, and the reflection will bend up or down, overcorrected or undercorrected. The velocity, if it is correct for the primary reflection, is not correct for the multiple. To smooth out the multiple requires use of the shallow velocity. So, when the primary lines up straight, the multiple will still have some NMO, and still curve downward.

Stacking—the combining of traces after correction—will weaken the multiple by combining energy on one trace with lack of energy on another, while the primary is combined energy to energy.

Although the velocity is not correct for the multiple, stacking won't just happen to make it exactly cancel, so multiples are weakened, attenuated, rather than eliminated, on record sections.

Filtering

A seismic source sends out sound of many frequencies. Some are audible, so you can hear the pop, or explosion, or whatever. Seismic energy, the sound that travels well through the earth, and thus carries seismic information is only a narrow range of frequencies, from about 10 to 100 cycles per second, cps, or Hertz, Hz.

The frequencies that do not carry seismic information, will clutter up the sections with extraneous noises—grass blades banging against the geophones, and things like that. Much of this unwanted information can be eliminated just by cutting out the frequencies that do not penetrate far into the earth. This is called filtering, or bandpass filtering, as it allows a *band* of frequencies to *pass*, but not others.

In analog recording, as in an anti-alias filter preceding digital recording, filtering is done electrically. An electrical arrangement is used, that discriminates strongly against some frequencies, and allows others to pass. A filter curve shows the percentages of the different frequencies passed.

With digital processing, filtering is more exact, with cutoffs as sharp as desired. Filtering digitally is an arithmetical operation, as is any digital treatment of records. The frequencies desired can be quite precisely selected, and the result will be just that. A special case is the elimination of power line frequency, 50 or 60 Hz, whichever is used in the country in which the shooting is done. In shooting on land, near an electric power line, especially if the ground is wet, induced or conducted electrical energy is picked up by the geophone cable. So, a "notched" filter is used. This is a filter especially designed to eliminate just that frequency and not others. Notice that in this case the unwanted energy has entered the system as electricity, not as sound later converted to electricity.

Data processors tend to call any operation on the data filtering, so it is sometimes necessary to specify bandpass filtering in conversation. However, the expressions filter test, filtered tape, and time variant filtering are all assumed to refer only to bandpass filtering.

Frequency Display

One of the early steps in data processing for a seismic program is to determine what frequencies of sound to retain for the seismic sections. A filter test, or frequency display, is made (Fig. 5-14). One record, or short piece of a section, is played out several times on a piece of paper, each time with a different narrow filter. One will have only some high frequencies, then in steps, the others will have lower and lower frequencies.

The display is examined to see which frequencies show some reflection best, then other reflections. High frequencies do not go deep into the earth, so the higher frequency records will tend to not have deep reflections, but will have the shallow ones. Selections are made, of the frequencies that show the data at a certain time on the records, the ones that show reflections at another time, etc. Other frequencies are eliminated from the final section.

If a few records from different parts of the area are thus tested, filters can be selected for the entire program.

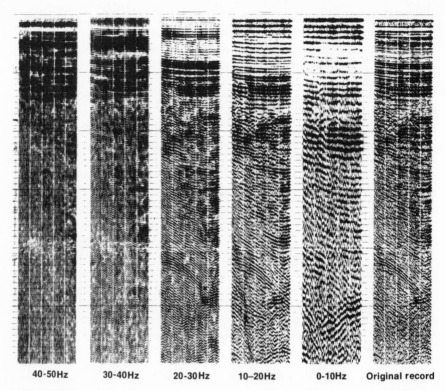

| 40-50Hz | 30-40Hz | 20-30Hz | 10-20Hz | 0-10Hz | Original record |

Fig. 5-14 Frequency display   Credit: Western Geophysical Co.

TVF

From the results of the frequency displays, a TVF, or time variant filter, can be applied to the data. A time variant filter passes a band of frequencies, but with the band changing, in maybe three or four steps, from top to bottom of the record. The bands are kept fairly wide, to retain any frequencies that contribute to a reflection. The bands are usually described on the labeling of the record section, either graphically or in tabular form.

Muting

In modern offshore seismic work, the farther geophone groups are quite far from the energy source, usually 2,400 ft. For shallow beds, refraction data crosses and obliterates some of the reflection information. However, shorter traces are not so affected. Therefore, in making a record section, the farther traces are muted—cut off—down to a depth at which reflections are free of refractions (Fig. 5-15). The farthest trace is muted to a greater depth than the next, and so on. So above the muting depth of the farthest trace the stack is not a full 2,400%, or whatever was the degree of coverage shot. Often, for some shallower beds, the degree of stack may be as low as say, 300%.

Then, for a different reason, the deeper parts of the nearer traces are sometimes cut off. These traces, near the boat motors, are noisier than others. The ratio of signal to noise is higher for the shallow data, where seismic signal is stronger, than for deeper data, with weak seismic energy but the same boat noise. For this deeper part of the section, farther traces, from geophones farther from the boat noise, yield better data.

Thus a 2400% record section may attain 2400% for only a medium depth, with far traces muted in the upper part of the section, and near traces in the deeper part. This, of course, is a part of the effort to make the data as good as possible.

In connection with this, it is important, when hiring a seismic crew, to thoroughly inform the contractor as to the depths of data most necessary. Streamer length and geophone configuration should be adjusted to depth of information sought. Thus, it is not enough to say you want the best seismic sections possible. The best possible shooting technique for a shallow basin will not also be best for a deep basin with a deep pay horizon. Still different is a deep basin with potential pay zones scattered up and down the section, which will call for compromise technique, not quite the best for either shallow or deep data.

Velocity Analysis

The traces of a CDP gather are combined to make one stacked trace. To combine them so the reflections are enhanced, normal moveout is removed. To do this, the farther traces must be pulled upward by varied amounts. That is, upper parts are pulled up more than deeper parts, to correct the different reflections' normal moveout curves, as they are less curved with increasing depth. (Fig. 5-16).

The simplest way to make this correction by computer is by trial and error. Apply different amounts of correction, and see which amount straightens a reflection.

There is a complicated equation for the relationship between velocity and normal moveout. Its essence is that a slower velocity produces

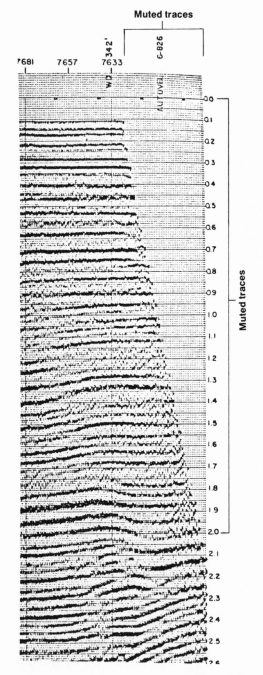

Fig. 5-15 Muting (detail of Fig. 12-1)   Credit: Teledyne Exploration

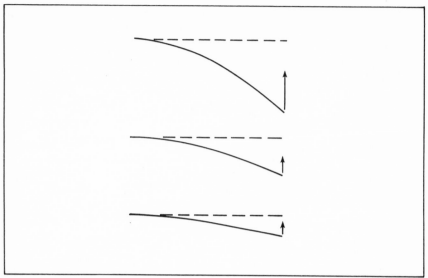

Fig. 5-16 Normal moveout corrections

greater normal moveout. With the equation, the trial and error corrections can be made in the form of corrections for normal moveout at specific velocities.

The CDP gather can be corrected for NMO at an arbitrary velocity, say 10,000 ft/sec. The correction will not make all the reflections line up with zero normal moveout, but will likely be correct for one reflection time on the record. At that point, a reflection will have no NMO, so the reflection will be aligned straight across the record. Other reflections will be insufficiently corrected, with some NMO remaining, so will curve downward; or over-corrected and will curve upward. So a look at the record will tell where that velocity applies.

If the same record is also displayed with normal moveout corrected at a different velocity, like 5,000 ft/sec, you can tell by inspection where the 5,000-ft/sec velocity is correct. So the record can be corrected at a number of different velocities, and an interpretation made to determine the velocities at the various reflections. The result can be plotted on a chart of time vs. velocity.

Exact velocities are difficult to obtain, as a reflection that is almost correctly adjusted will look about as straight as one that is correct. So some other techniques have been developed to help make this fine decision. One such technique is to stack all the traces of the gather, corrected at some velocity, to make one composite trace. Then the reflection that is best correlated will produce a higher-amplitude event than another. Each of the corrected records from the different velocities can be so stacked, and the resulting traces put side by side like a record. Looking at this display, a better determination of the best velocity can be made.

Other, more sophisticated, ways of detecting the best velocity are in use. They involve computer detection of amplitudes, coherence, etc., and are displayed in various ways, with horizontal traces through the velocities (Fig. 5-17) rather than vertical for each velocity, or with contours around amplitudes instead of traces, etc.

Often, as a supplement to one of these displays, a stack of a short segment of the line is shown, at the various velocities. So there is a piece of section stacked using 5,000 ft/sec, the same piece at 5,500, etc. Looking at a horizon on the piece of section that shows it best, not only indicates the approximate velocity for that reflection in terms a geophysicist is accustomed to, but also shows what the final product will be like.

With this kind of velocity information, several things can be done. For one, a number of analyses can be run at points along the line, and used for normal moveout correction so traces can be stacked. The same information is an aid in converting times to depth. In areas of strong lateral velocity gradient, this sometimes makes the difference between false and real structure.

The normal moveout velocity data is poorer for deep reflections, as there is less normal moveout to work with, but, to exactly the same degree, there is less normal moveout to be corrected for, so the correc-

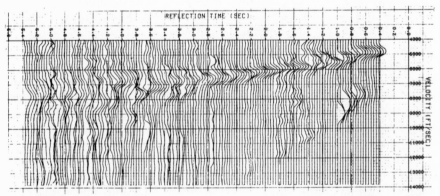

Fig. 5-17 Velocity analysis   Credit: Western Geophysical Co.

tion works just as well at greater depths, for this one purpose of producing a better stack. But the deeper data is not as good for other purposes.

A new use of the data, developing as velocity determinations improve, is to identify horizons by velocity. For instance, a fault may break a horizon and the reflection may not be recognizable enough to correlate across the fault. In this case, a certain velocity found on each side of the fault will sometimes help decide how a reflection should be correlated across the fault. Actually, the velocities on the two sides might not be quite the same, as greater overburden on the downthrown side would tend to make the velocity somewhat greater. And an even greater difference would probably result from the imprecision of the velocity determinations.

This leads naturally into another use that is being developed, the recognition of lithology by velocity. As an extreme example, a carbonate may stand out in contrast to a shale on the basis of velocity. And in the sand and shale zones, the overall sand-shale ratio can be roughly determined by velocity.

This lithology detection is, like the recognition of horizons, in a beginning stage. Velocities from normal moveout are very rough approximations. They are as much more crude than a seismic record as a record is cruder than a well log. So, not much can yet be done along these lines, and often nothing clear will result from the velocities. Yet it is an exciting idea, and will surely improve with time.

Another use of seismic velocity information is the detection by velocity of high-pressure zones, to aid engineers in planning mud programs, etc., for drilling wells.

One form of display of velocity data is as a cross section. Velocities may be plotted as curves where the analyses are made, or lines across the section can be plotted, tracing zones of equal velocity, or both (Fig. 5-18). The display can be made on clear film at the same scale as the record section, and overlaid on it to show the relationship between reflections and velocity.

The commonly used name for velocities obtained from normal moveout data is "RMS velocity". This stands for Root Mean Square, a mathematical curve-fitting method that may be used in deriving the velocities.

Other ways of determining velocity information, from wells rather than from record sections, are described in the next chapter, Chapter 7, Data Handling.

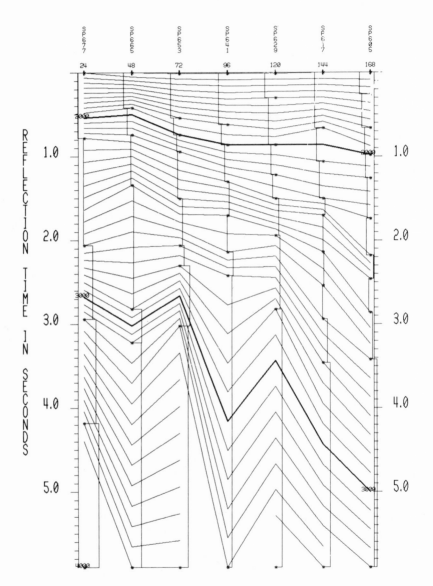

Fig. 5-18 Interval velocity curves and equal velocity lines

Credit: Bruce Meadours Software Services

Signal Theory

Digital seismic processing uses techniques from signal theory, which was originally developed to improve radio and radar. It is mostly concerned with clearing up the message received, and reducing interfering noises.

Its main techniques involve combining two lists of numbers, to obtain a third list, that bears some desired relationship to the two. This is mostly done by multiplying numbers in one list by those in the other, and is, in practical use, done by computer.

Various things can be done with this basic technique. One list can be reversed before shifting past the other. One list can be a specially designed one, an "operator", usually shorter than the other, and chosen to produce a third list that is some desired modification of the one operated on, the longer list.

The lists of numbers can be plotted as curves, and can be numerical representations of data received. So they can be seismic traces in number form, which is the form of seismic data obtained in digital recording.

So operations using signal theory are some of the means used to improve data in digital processing. They include correlation, convolution, digital filtering.

Correlation

Correlation is one of the applications of signal theory to seismic data.

The wiggly line that is a trace on a record section, is in digital form a list of numbers, so the signal theory techniques can be applied to it. The numbers are samplings of the data, every one, two, or four milliseconds. The wiggles to one side and the other of the central position are represented by positive and negative numbers.

Putting two of the traces side by side in digital form, there are two columns of numbers. If the first number of one trace is multiplied by the first number of the other trace, then the second by the second, etc., a column of products is the result. Adding the products together yields a single number, which is a measure of how much alike the two traces are, in that relative position. The bigger the number, the more they are alike.

Multiplying, rather than adding the pairs of numbers, makes matching peaks and matching troughs both yield large positive numbers, as the product of two positive numbers is positive, and the product of two negative numbers is also positive. A peak matched with a trough though, produces a negative number, as the product of a positive and negative number is negative. So a good match is positive, and a mismatch is negative.

Then, if one trace (list) is shifted with respect to the other and the whole process repeated, a new sum of products is obtained. Then shift further, and get another, etc. Then list these sums of products, as the trace resulting from the correlation operation.

Plotting the third list as a wiggly line trace makes a visible display of how much alike the two traces are in their different positions. There will be an amount of shift that makes the best fit, having the largest sum. A shift either way from there will produce poorer fits, and smaller sums. And with the seismic traces composed of alternating peaks and troughs, the best fit will be at a shift that puts the traces peak to peak and trough to trough. Then, a shift to a peak to trough match will have a very poor correlation (negative sum), and a shift to the next peak to peak position will yield a better correlation (positive sum), but not as good as the best fit. So the plot of correlating sums will itself have peaks and troughs (Fig. 5-19) as the correlation gets poorer away from the best fit position. A correlation display is a correlogram.

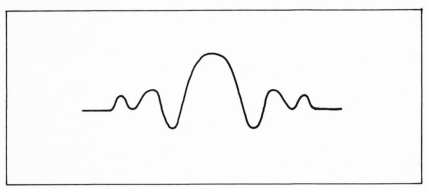

Fig. 5-19 Correlation plot

This, then, is a digital technique that can tell how much alike two traces are, and how they must be shifted to match best. If the two traces are different, the process is called cross-correlation.

Correlograms are trace-type curves, with a big peak at the correct correlation, and lesser peaks at the distances the other good correlations are apart. As a shift in either direction from the best fit shows exactly the same features, the curve is symmetrical (Fig. 5-20). There is no need to show the two identical halves, so the correlogram displayed is usually only half the curve, (Fig. 5-21).

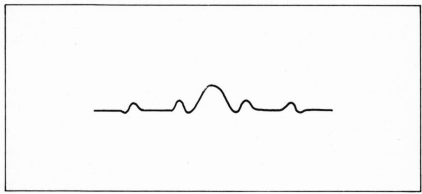

Fig. 5-20 Correlogram is symmetrical

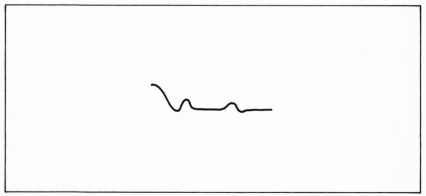

Fig. 5-21 Half of correlogram

A special case of correlation, and at present the most useful one, is auto-correlation. As the name implies, it consists of correlating a trace with itself. That is, the two traces correlated are copies of the same one. The same procedure is used, shifting one past the other, multiplying, adding products, listing, and plotting sums. Now obviously, the best match will be when there is no shift. The traces then match perfectly. It hardly takes a computer to demonstrate that a trace matches itself with no shift. But the interest here is in any other fairly good correlations that may show up as strong peaks on the auto-correlation trace, the auto-correlogram.

If a trace can be shifted some amount, and *then* correlates well with itself, then it must have some repeating feature. For instance, if the original trace has a strong reflection every 165 milliseconds, then, when the two traces are shifted 165 ms from one good fit, they will fit fairly well again. So the auto-correlogram will have a strong peak 165 ms from the best fit peak (Fig. 5-22).

Auto-correlograms, one from each trace, can be displayed side by side like a record section, and can be shown with the normal section, to aid in interpreting whatever multiples, etc. could not be removed, the remainder after the attenuation produced by the processing.

Deconvolution is a process to minimize the repetitions on a trace, so auto-correlation can be used to determine the kind of deconvolution needed, and the effectiveness of deconvolution after it has been applied.

So a ghost, or the several parts of the wavelet, or a multiple, being repeated through the trace, will be evident on the auto-correlogram (Fig. 5-23).

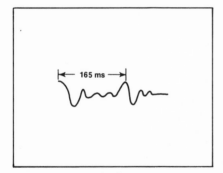

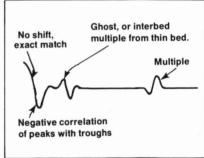

Fig. 5-22 Auto-correlogram   Fig. 5-23 Features of auto-correlogram

Convolution

A process similar to correlation is convolution. It is performed with two traces, shifting, multiplying, adding products, and plotting the sums, just the same as in correlation, but with one of the traces reversed, end for end.

The effect of convolution is not to determine how much alike the two traces are, but to produce a trace that is a repeating combination of the two, the way a seismic signal combines with many velocity interfaces in the earth to produce a seismic trace. In fact, the seismic signal is spoken of as being convolved with the earth.

To see how this works, consider a long "trace" consisting of a straight line, with a perpendicular bar at the time of each velocity interface in the ground (Fig. 5-24). The bar, "spike" in signal theory

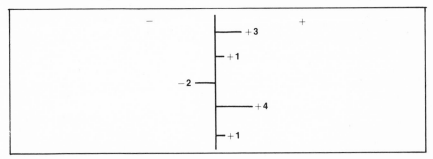

Fig. 5-24 Interfaces as spikes

language, has its length proportional to the reflectivity of the interface, that is, the amplitude of the reflection it will produce. And the bar is to the right or left, positive or negative, according to whether the velocity change is from slow to fast or fast to slow. This trace represents the velocity interfaces in the subsurface, and is put in numerical form, all points on the straight line being zero, and the spikes, single positive or negative numbers, (Fig. 5-25).

0	0	0	0	+3	0	0	0	
0	0	0	+1	0	0	0	0	etc.

Fig. 5-25 Spikes as numbers

Then another trace, a short one consisting of just the seismic wavelet, in the form of a list of numbers, is reversed and moved past the long trace (Fig. 5-26). Assume that the first spike on the long trace is isolated, with no other spikes near it. In any position of the wavelet in passing it, all the products are zero (zero multiplied by anything is zero, or none of anything is nothing), except for the one at the spike, which is the spike value times the part of the wavelet at that point (Fig. 5-27). So the list of products is the product of the spike and the first

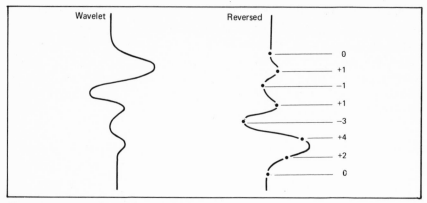

Fig. 5-26 Reversed wavelet as numbers

First position								
0	0	0	0	0	0	+3	0	spike
+1	−1	+1	−3	+1	+4	+2	−	wavelet
0	0	0	0	0	0	+6	−	product
Second position								
0	0	0	0	0	0	+3	0	spike
−	+1	−1	+1	−3	+1	+4	+2	wavelet
−	0	0	0	0	0	+12	0	product

Fig. 5-27 Wavelet passing spike

point of the wavelet, the spike and the second, etc., (Fig. 5-28). So the plot of sums of products is a plot of the wavelet, but no longer reversed, and multiplied by the spike. So the wavelet is larger or smaller, as the value of the spike is a number greater than one, or a fraction. And, if the

+6	+12		−9	+3	−3	+3

Fig. 5-28 Spike-wavelet products

spike is negative, then the wavelet is of opposite polarity, a mirror image with left and right interchanged.

Then when the short trace arrives at the next spike, the wavelet is plotted again, larger or smaller, positive or negative, and so on for all the spikes. Where two spikes are close together, the wavelets will be mixed together, and not individually recognizable on the resulting trace.

This is what the seismic wavelet does in the earth. The first part of it arrives at the uppermost reflector, then the later parts of the wavelet arrive at the reflector. Later it arrives at the next reflector, etc. The reflections going back up arrive at the geophones in the same sequence, the first part of the wavelet bounced from the top reflector, etc. And the wavelets of close-together reflectors are combined into more complex shapes, (Fig. 5-29).

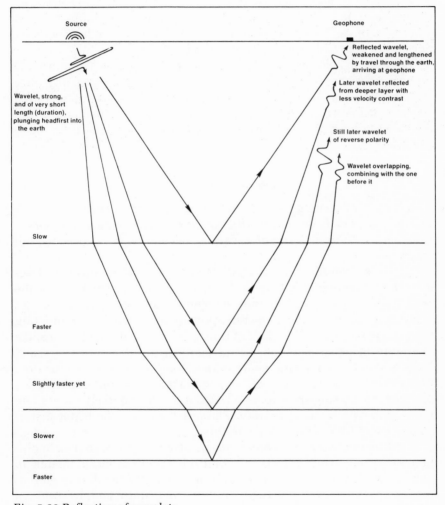

Fig. 5-29 *Reflection of wavelets*

Therefore, in imitation of nature, a similar procedure is used in making a synthetic seismogram from a sonic log and velocity survey in a well. The well data provides information to make a line with spikes representing the subsurface. Times to velocity interfaces and reflectivities at the interfaces are calculated. Then a wavelet like one that would be produced by a seismic source, is convolved (reversed and correlated) with the spikes. The trace formed by the convolution is a fabricated, or synthesized, seismic trace with all reflections definitely identified. It is usually made to be even more like an actual trace by adding the effect of density, and including calculated multiples and similar features. This process is described more fully under Synthetic Seismogram, in Chapter 7, Data Handling.

Convolution also has more general uses in processing. The shorter of the two traces convolved need not be a wavelet but may be any trace that, in being convolved with another trace, modifies it in some desired way, or operates on it. So the short trace is called an "operator". And the long trace can be a regular seismic trace that needs to be changed in some way.

Deconvolution

When seismic energy is initiated, it all occurs at once, as a near-instantaneous bang or pop. But the sound received by the geophones is stretched out into the form of a wavelet, several wiggles stretched out over some tens of milliseconds.

With this stretching of reflected energy, some reflections can interfere with others nearby. One is still going on when it is time for the next, so where they overlap, they are blended for some milliseconds.

Deconvolution is a mathematical process for partly re-compressing the stretched-out wavelet into a shorter one—not completely to a spike of energy like that at the start, but more nearly like it. The trace is processed with an operator, another trace, that, combined with the first, has the effect of reducing repetitions in the trace. The different wiggles of a wavelet are squeezed up, so a reflection is not so likely to interfere with a following one (Fig. 5-30). The reflections can be better distinguished, and picked as separate events.

Ghosting and water reverberations are also repetitive, so deconvolution can reduce their effects too. Multiples that are close enough to their primaries, are also diminished. In doing all this, deconvolution tends to make sections obtained with different equipment, by different

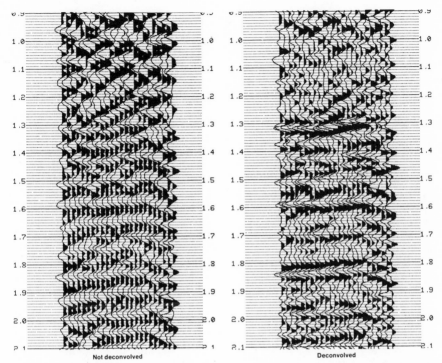

Fig. 5-30 Effect of deconvolution   Credit: Houston Oil & Minerals Corp.

crews, look more alike, so they can be incorporated into a single interpretation more readily.

Selection of the right operator by the processors requires fairly good records, so poor records, with a lot of noise, can't be deconvolved as well. And deconvolution doesn't do anything to eliminate most noise. It works on each trace separately, but reduction of most kinds of noise calls for a method that works on several traces together.

Deconvolution, or decon, is normally performed on seismic data before it is stacked, that is, on the unstacked traces. This is DBS, decon before stack. Then, the stacked traces may also be deconvolved. Tests are often run to see if this second deconvolution, DAS, decon after stack, will improve the data. If it does not seem to help, then it may be left out, to avoid unnecessary tampering with good data.

Signature deconvolution enhances the data still more. It can be used when the signature of the source was recorded with each shot, pop, etc., by a geophone near the source. These wavelet shapes are then used to make the deconvolution process better fit the actual data (Fig. 5-31).

Fourier and Frequency Domain

If two simple sine (or cosine) waves, each having uniform amplitude and frequency, are combined, the resulting wave is more complex than either. Adding more, a wave of any desired complexity can be obtained.

Conversely, any wave shape, like a seismic trace, can be duplicated by a combination of such waves, using just right frequencies, amplitudes, phases. Phase is shift in seismic time, or where the starting point is on the curve.

A "Fourier analysis" determines mathematically what waves will combine to duplicate the trace. Each of these waves can be defined completely by giving frequency, amplitude, phase. So the seismic trace can be described by giving the frequencies, amplitudes, phases of all the waves that make it up.

This is a different way of looking at the information in a trace. For instance, it tells more about the frequency content of a trace, and less about the time relationships in it.

A normal trace is said to be in the "time domain". The Fourier analysis puts the trace into the "frequency domain". The value of this is in having a different way of handling the data in the trace. Some data

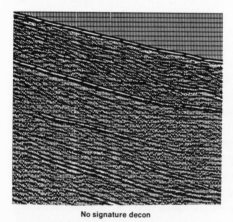

No signature decon

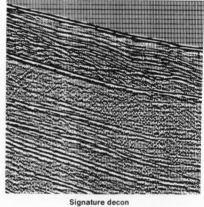

Signature decon

Fig. 5-31 Signature deconvolution   *Credit: Seiscom-Delta Inc.*

processing is more readily handled in the time domain, some in the frequency domain.

Frequency domain data can be displayed as two graphs, one of frequency vs. amplitude, the other frequency vs. phase.

The frequency vs. amplitude plot is useful in that it shows, in an easily-understood way, the frequency content of the trace. A "power spectrum" is a variant of it usually produced as a computer printout. It is frequency vs. the square of amplitude. It conveys the same information.

If a trace is in the frequency domain form, some frequencies can be easily and cleanly removed, by just eliminating that part of the data. Then, when put back into the time domain, it does not contain the unwanted frequencies. This is bandpass filtering in the frequency domain.

Flattening

A process occasionally used, especially in areas of difficult near-surface corrections, is flattening a horizon.

The main purpose of flattening is to get around all the near-surface problems, and produce a record section that shows reflections and their relationships with each other quite clearly. Especially in areas where low-velocity glacial drift accumulates unevenly near the surface, the near-surface corrections can be a slow, unreliable, and unnecessary chore. With a reflection flattened, the uncertainties and residual irregularities are eliminated, and considerable time is saved. Drape over a reef, relative to other beds, will detect a reef, regardless of other structure. The other structure can influence the situation, of course, and a structural section may be advisable before drilling. The precision of the structural section, though, is not so critical in this back-up use. For flattening, a good reflection should be selected. It can be almost anywhere in the section, and should be selected primarily for quality. The only real purpose it has is to make easy and accurate the observation and timing of divergences between horizons that indicate drape over a reef, or whatever similar phenomenon is sought.

A quite different reason for flattening a horizon is to restore the section to some particular geologic time—after deposition of some formation, for example, assuming that the deposition was flat. In this case, the selection of a horizon to be flattened is done on a geological basis. It should be the level desired, if that is pickable. If not, then the nearest reliable pick expected or appearing to conform to the desired formation can be flattened. If the flattening of the particular horizon is difficult, it may be helpful to first flatten whatever is the best pick on the section. Then the flattening of the desired horizon may be easier, with this smoother section to work from.

It may be useful to make several sections, with various horizons flat. They can then represent a sequence of geologic times, to be used in a study of the geological history of the area.

Dip Rejection

Dip rejection, also called a moveout filter or velocity filter, removes selected moveouts on the section, that is, selected dips. It has two main uses. First, in an area of rather flat bedding, steep dips can be removed, to clean up the section. Much of the random noise will fit the computer's criteria for steep dip, as a wiggle on one trace will have some milliseconds of moveout (difference in time) separating it from a nearby wiggle on the next trace. If you know there aren't steep dips, the machine can throw out anything that looks like a steep dip (Fig. 5-32).

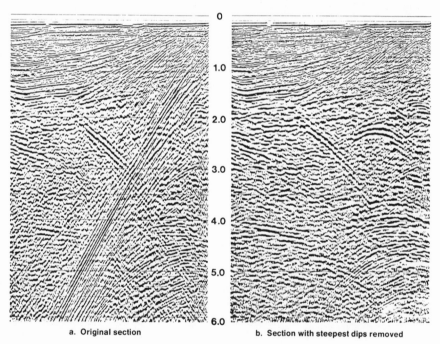

a. Original section   b. Section with steepest dips removed

Fig. 5-32 Dip rejection   Credit: Digicon Inc.

The second use is in taking out multiples, by removing certain dips. For instance, sometimes many flat layers occupy the shallow part of the section,and multiples of them may obscure steeper dips in the lower part of the section. So the section can be played back with flattish dips eliminated. This destroys the real information in the upper part of the section, so another copy of the section, without that treatment, is also needed (Fig. 5-33).

In both cases you are assuming you know what the section should show. This is, of course, dangerous as what you "know" may not be correct. So moveout filtering is sometimes helpful, but should be used with caution, and when used, viewed with some suspicion.

Processing Sequence

The processing of seismic data necessarily follows a fairly uniform series of steps. Different processors have their own variations of processing, and may vary the order in which the steps are performed, but they all have about the same processes to go through. The following are more or less the steps they go through, in more or less the order they use.

a. Original section

b. Section with flatter dips removed

Fig. 5-33 Flat dip rejection   *Credit: Digicon Inc.*

Processing naturally begins with the arrival in the processing center of the first batch of field tapes from the crew. They are accompanied by written records of line number, when shot, disturbances from extraneous noises, field problems, etc.

The field tape is edited. It is not itself changed, but the data is recorded in modified form on another tape. It is first demultiplexed, so each trace is separate, rather than intermixed with the others as on the field tape. Dead or poor traces are eliminated.

The gain recorded is taken into account, in one or both of two ways, either to produce a section that has the gain so balanced as to make all reflections look fairly good, or to make a section with "true" amplitude, with the strength of a wiggle on one trace truly comparable to the strength of one on another trace.

The data is deconvolved, compressing each wavelet into one with less vertical extent, so the various reflections tend not to interfere with each other so much.

Statics are corrected, moving the traces up or down to correct for elevation, weathering.

A sampling of records are analyzed for velocity. RMS velocities are obtained at these selected points.

Normal moveout is removed, using the RMS velocities, interpolated between the points at which they were obtained.

Traces are muted, eliminating shallow parts of the longer traces (farther from the shot), and perhaps deeper parts of short traces.

Residual statics may be applied, straightening out irregularities left by the original static corrections.

With these corrections made, the traces are ready to be combined into one trace for each depth point. They are stacked, and a stacked tape is produced.

Sometimes it is useful to deconvolve the data again after it has been stacked.

Filtering is applied, usually as time variant filtering, TVF, in which shallow data is restricted to certain frequencies and at other depths to other frequencies. A TVF tape is made. This is the final tape.

From the TVF tape, record sections are made. They are the final product of normal processing.

Tests. Before the processing is undertaken for the batch of data as a whole, some tests are made to determine the best way to do the processing—what filters, how much decon, decon before stack only or both before and after, the best looking amplitude setting, timing lines, scales, etc.

These tests are customarily shown to the client, and discussions held to establish how the data for the area is to be processed. The tests include frequency displays to decide which filter settings to use, examples of sections with varying deconvolution, etc.

Each section is also usually produced in a preliminary form, directly on paper, with no film, so the processors can check it before turning out a film section. This preliminary section gives the client too a chance to see what changes in processing may be necessary to fit his needs.

Routine. The routine for the processing is filled with entering data into the computer, having the computer print "data dumps", checking them, scheduling the stages of various clients' work so they all are kept progressing, and the computer does not have idle time. The main thing noticeable in a visit to a processor is people working with pieces of paper.

Special Processing. The foregoing is the normal processing, that is used on most sections. Special processing, performed in addition to the normal, is making migrated sections, sections converted to depth, sometimes sections with variations of the normal processing—maybe a different set of filters or different decon to compare with the other section. These special processes are in some cases performed after the other processing, sometimes fitted into the sequence.

Tape Storage

Magnetic recording tape for use in computers must fit much higher standards than home "hi-fi" music tape. There are requirements for close tolerances, so a bit, recorded on the tape, will be read by another machine as being in the same spot, and in the same relationship to other bits.

Any distortion of the tape, by stretch or shrink, will affect the precision of location of the bits. The backing of the tape is Mylar,* which is about as resistant to distortion as any material, but under highly adverse conditions, anything can be distorted.

Dust or dirt on the tape can affect the ability to read all the bits, the 1's and 0's. Magnetic or electrical fields can affect the magnetization. A strong enough magnetic field would erase it.

The tape, then, although it is tough, and the information on it is in clear-cut bits, has so much and so tightly packed information on it, that it is vulnerable to a number of things.

Before data is recorded on it, the tape is usually kept sealed in its original containers. After recording, the data on the tape needs to be protected from damage.

Data is recorded in the field, where the weather may be bitter cold,

---

*Dupont registered trademark

baking hot, very humid, dry and dusty. There may be fungus in the air, salt spray, sandstorms.

The tape should be, and is, fairly well protected during the recording. Recording is usually done in air conditioned instrument rooms or "doghouses," as most recording equipment requires a moderate temperature and humidity, for smooth operation. Tapes, after recording, are immediately resealed in plastic cannisters and plastic sacks. When a cardboard box that tapes come in is again full of tapes, it is sealed up and addressed to the processing center it is destined for.

The greatest risk to the tape is between the crew and processing center. Boxes of tape are sent from the crew. There may be only one shipment at the end of the operation or there may be several, especially if the program is a large one. When they are sent in, the tapes are assigned to the tender mercies of someone—shipping company, airline, client's crew boat, somebody with a vehicle who is going that way. At this stage they are probably in the care of someone who has other concerns, more important to him than nursing tapes, and who does not know what conditions will harm them. They should not be:
Left out in the sun,
Left out in the cold,
Left out in the rain,
Carried on deck in salt spray,
Stacked near heating pipes,
Stacked under heavy weights,
Thrown around,
Checked by magnetic detectors
or x-rays,
Placed beside electric motors or dynamos,
Shipped with magnetic materials,
Delayed long in reaching their
destination.

All of that is too much to keep track of, so for some of it you have to depend on luck, but the tapes are the entire result of the expense of the seismic survey, so it's worthwhile to put forth some effort to impress people with the need for extra care.

Signs like "Magnetic Tapes—Do Not X-Ray" can be put on the boxes. Insuring for a high value impresses shipping companies.

In the processing center, tapes, being the center's life blood, are well cared for.

After processing, the tapes that are to be kept are best preserved by a tape storage company. When stored, the storage company should be asked to "clean and tension wind" the tapes. The tension winding makes the tape uniformly tight all around the reel, eliminating one cause of distortion. The cleaning and tension winding should be repeated about once a year. The storage company will maintain an optimum temperature and humidity, and can index the tapes, get requested reels out of storage for reprocessing, etc.

If for some reason it is not practical to leave the tapes in a specialized tape storage, and they are kept on the company's premises, then a simple rule of thumb is, keep them in an air-conditioned room. The temperature and humidity that are most pleasant to people, are also good for tapes. Why should magnetic tape be so like people? I suspect it is because the development of magnetic tape recording took place in comfortably air-conditioned laboratories. The things that experimentation showed to work best, were those that work best in those conditions. If all the inventing along that line had been done by a guy with poor circulation, who kept his lab fairly hot, you can bet that the tape and recording equipment would be somewhat different, and that people in processing centers would hate the hot rooms they had to work in, that were required for magnetic tape and tape recorders. So we're lucky. A visit to a tape storage is pleasant.

6

Record Sections

Record sections were described in Chapter 1, to get things started. Here, we'll go over them again, in the context of processing and as derived from individual records.

The record of a single shot does not become a distinguishable part of the record section. Instead, it is taken apart into traces, and the traces modified and combined with traces from other shots, to form stacked traces in the section. You could say that the, say 24, traces from a shot become parts of 24 consecutive traces in the final section, but that each one is only a 24th of the trace it is a part of (Fig. 6-1).

The section then, is made up of many stacked traces. Normal move-out has been removed in correcting for stack, so the traces are all as though recorded by a geophone right by the source.

The timing scale of a section is uniform, in seconds, and is vertical on the section, as are the traces. The horizontal scale, though, is not always uniform. Traces are spaced uniform distances apart. This can be set at any of a number of spacings of so many traces to the inch. However, the traces represent positions on the line, and the line is assumed to be straight. If the line bends, or if the pops in offshore work are not as uniformly spaced as intended, or if on land physical obstacles cause the spacings to be non-uniform, then the traces will still be uniformly spaced on the section, but those spaces will not represent equal distances on the ground. This is generally a small discrepancy, but does indicate that the way to get true angles and true distances from a section is to read shot point numbers from the section, and refer to the shot point locations on a map.

Offshore sections are either adjusted so the zero time on the section is at sea level, or left unadjusted, so zero is at an average of source and geophone depths, around 30 feet under water. A distance below the zero line, will be the first reflected energy. This is easy to consider as

132

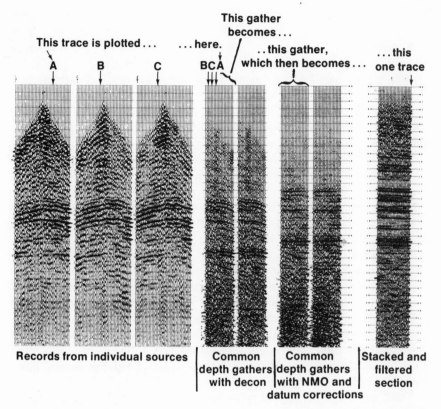

This trace is plotted... ... **here.**

This gather becomes...

..this gather, which then becomes...

...this one trace

A   B   C   BCA

Records from individual sources | **Common depth gathers with decon** | **Common depth gathers with NMO and datum corrections** | **Stacked and filtered section**

Fig. 6-1 Steps in processing   *Credit: Houston Oil & Minerals Corp.*

being the sea bed. However, where the bottom is soft and water saturated, there may not be much velocity contrast between water and mud, therefore little or no reflection. In this case, the first reflection seen will be the first hard substance below the bottom.

Records made on land are, in processing, adjusted to sea level. In some processing, the hills and valleys appear as hills and valleys on the upper ends of the traces. In other situations, the near surface parts of the section are handled differently, so no such direct relationship is visible. Often the surveyed elevations are plotted along the top of the section. This is useful for showing the effect shooting on high or low places has on reflection quality in the particular area, and for making multiples more apparent, as mirror images of the topography.

Display

A seismic section is most often made up of traces in the form of wiggly lines, with the peaks filled in, but there are also other forms in which traces can be displayed. Each has its own advantages and disadvantages.

A wiggle trace section has its traces just as wiggly lines alone. They are close together, and overlap where there are strong reflections. The section is fairly easy to work on at a desk, marking the lineups of reflections. Even though there is the overlap, there is no blanking out of data, so the fine detail of the wiggles can be best seen on this form. However, such a section is not so good when looked at from a distance, either so the interpreter can get an overall view, or in a meeting, to show it to a number of people. Then the traces optically blend, so the section just looks grey all over.

A variable density section gets away from the overlap of traces. In it, each trace is a narrow band of varying shades. Dark grading to white replaces the variation from peak to trough. The band is straight, and of uniform width. The variable density section is easy to observe from a distance, with the reflections showing as light and dark lines across the section. However, they are not so easy to handle at a desk. Where a trace joins the next one, the slight differences in times of a reflection show up as sharp discontinuities. If there is steep dip, the section has a stairstep look.

The tone range of grays, from white to black, cannot be reproduced with the high contrast diazo (ammonia developed) prints usually used for record sections. For viewing from across a room, this loss of detail is no problem, and may make the image appear more distinct, but the lost detail makes a precise picking impossible.

Variable area is another type of display that is easy to see from a distance. It is based on the wiggle idea, but doesn't have the wiggles. It consists of the peaks only, greater than some chosen amplitude, filled in solid. From a distance, reflections stand out even more clearly than they do on variable density. But, working at a desk, it is apparent that most of the detail is gone. There are no troughs, and where the peaks overlap, the solid color blanks out their very tips. So neither bottoms of troughs nor tips of peaks can be picked. This also takes the precision out of picking.

A way to combine advantages and minimize disadvantages is to combine two kinds of display, one for precision picking, the other for visibility at a distance.

A combination of wiggle and V.D., variable density, is occasionally used, but is not very popular.

Wiggle and V.A., variable area, is a natural. The variable area fills in part of the wiggle, so seems to be part of it, and makes the reflections stand out. There isn't a loss of wiggle trace detail, except where peaks overlap. The bottoms of troughs can be picked precisely. There is enough white space in the troughs for marking the section with colored pencils. So this is the predominant form of record section, to the extent that, when a section is mentioned, people think of a V.A. plus wiggle section.

Color

Seismic sections are appearing in color, more and more. They are in the form of photographic color prints, which makes them expensive and calls for a color printing setup. The expense tends to limit their use to special investigations. And the color printing can't be done conveniently on the office blue line printer.

The color sections are usually normal wiggle plus variable area, with some other parameter added in color. Velocities may be shown as gradations from red through the spectrum to blue. Frequencies or polarities may be shown by color. Even amplitude or dip, already visible on the section, may be given added stress with a range of colors.

The color sections put more data in usable form on the sections, for direct comparison with the usual information. For instance, a person can see and think about velocity variations while he is looking at structure, and better understand the interrelationships between the two.

Vertical Exaggeration

Seismic sections do not, strictly speaking, have a ratio of horizontal to vertical scale, as the horizontal scale on the section is a matter of distance, while the vertical scale is in terms of seismic time.

However, there is certainly an *apparent* relationship of vertical to horizontal scale on a section. People show sections, and point out places with steep dip, etc. It is natural to think of both vertical and horizontal scales, on a display that is as pictorial as a seismic section, as being the same kind of thing. Thought of in this way, there is a vertical exaggeration, of varying degree. A tenth of a second may in the shallow part of the section represent a vertical distance of 400 ft, while the same interval in the deeper part might cover 1000 ft.

In some respects, it would be more convenient to convert the vertical scales of all seismic sections to feet, and make vertical and horizontal scales the same. Then all dips and fault angles could be determined by protractor. Structures would not look misleadingly steep. There would be many advantages to a geologist looking at the data.

Sections are indeed occasionally converted to depth (Fig. 6-2), and sometimes even to a 1:1 scale. But there are a number of reasons why this is not practical for general use.

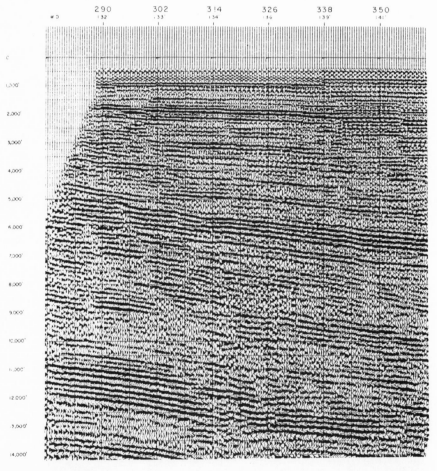

Fig. 6-2 Depth section   *Credit: Western Geophysical Co.*

Although the determination of RMS velocity from seismic data is becoming more definite with improvements in data processing, it is not yet very good for determining depth to horizons. And the deeper the horizon, the less reliable the depth determination. Velocity obtained from a survey in a well is not reliable at a distance from the well— sometimes a mile away is too far to trust the velocity. So there isn't, in general, a dependable way to convert time to depth.

Seismic time is the thing that is measured, and the farther we stray from using time the less definite the information is, so conclusions are on shakier ground.

Converting sections to depth is also an expensive process, so, unless highly beneficial, should not be performed on all sections indiscriminately.

The apparent vertical exaggeration allows subtle features to be seen more clearly, so makes interpretation easier. In fact, the need for even more horizontal compression is evident in the tendency of a geophysicist to look at a section from the end, with his eye down close to the section and squinting along it. This produces a foreshortening that makes it possible to determine the reflection lineups a little more confidently.

There is a storage and paper handling problem also. If seismic sections were all stretched out horizontally to 1:1 scale, while retaining a vertical scale large enough to show necessary detail, considerably more storage space would be needed, and more unfolding or unrolling of sections in working with them.

So conversion of sections to depth, and to 1:1 scale, are techniques to be used judiciously, in the cases in which they will serve a special purpose for interpretation or display. As RMS velocity information is obtained from reflections, conversion to depth should be reserved for sections having some very good, strong reflections to base the velocity information on.

Compressed Section

A greater vertical exaggeration than usual for a seismic section can be produced by making the section with its horizontal scale compressed. This can be done in processing, by putting the traces closer together, maybe combining or omitting some traces. Or it can be done photographically. The photographic method requires a special camera that will reduce in one dimension but not in the other.

A compressed section is made with considerable compression, say

10 to 20 to one, so the vertical exaggeration is really extreme. The main advantage of the compression is to make otherwise subtle features easier to discern. It is used along with the normal section, to help in picking the regular one.

A gentle, low relief structure that might go unnoticed on the regular section, may have obvious curvature on the compressed section. Faulting that is apparently very low angle, will appear more vertical, and therefore perhaps easier to recognize.

Timing of a reflection, at a specific shot point or other surface point, in order to be located on a map, is more difficult on a compressed section. It is harder to determine a point that is straight below some point on the surface, as the eye or a straightedge can lean to one side. So the regular section should be the one on which reflection times are picked for transfer to maps.

Sheer compactness of the compressed section is another of its advantages. If several special types of processing are used on a line, it may be more workable, in terms of paper handling for comparison, and of room on a desk or wall, to have some of the sections in compressed form.

Such things as the hyperbolic shape of diffractions and the similarity, as a mirror image, of multiples to the surface topography may be clearer on a compressed section.

A compressed section, if made by processors, requires about the same computer time in running tape that a normal section does, so it may cost about as much as any additional display. So for economic reasons, it is worthwhile to combine it with something else that is being done. That is, if some different display, with special filtering, or true amplitude, or something, is being made, then perhaps it can be made in the compressed form. Or some preliminary stage in the processing can be made in compressed form. If a photographic means of achieving the distortion is available, it will be considerably cheaper, and probably more flexible.

A compressed section, once made, should not be filed separately from the regular section, but made a part of it. The films can be spliced together, so any print of the section will include both. This can save filing effort, and manhours in hunting around the office for sections.

7

Data Handling

Velocity Determination

Velocity was discussed briefly in Chapter 1, to introduce the idea, and had a place in Chapter 6, Data Processing, in the special context of obtaining velocity information from seismic data. Now we need to look at it from the viewpoint of the interpreter, for converting seismic times to depth, and identifying horizons.

First, all velocity calculations use one simple basic formula—velocity times time equals distance, $VT = D$. It's so basic that everyone using it should remember it without thinking. I can't, though, so I say to myself, 40 miles an hour for 2 hours takes you 80 miles. Then I write $VT = D$, and proceed to use it. That's all there is. Multiply time by velocity to get distance, or divide distance by time to get velocity, $V = D/T$. Or divide distance by velocity to get time, $T = D/V$.

Here is an example. Assume a velocity of 10,000 ft/sec, and a seismic reflection at 1.231 sec. How deep is the formation? O.K., multiply 10,000 by 1.231, and you get 12,310 ft. But, the sound went down to the formation *and back*, while you just want to know how far it is down *to* the formation. So you have to divide the answer by 2, to get the depth. That's better, 6,155 ft.

Now if 10,000 ft/sec is a good velocity for that horizon in that area, you'll be multiplying a lot of reflection times by 10,000, and dividing by two. So to save time and reduce errors, you might as well multiply by 5,000, doing both operations in one step.

This using half the velocity to keep from having to divide by two is commonly used, and sometimes leads geophysicists to talk about half-velocities. But although the whole velocity is rarely used in calculations, it is almost always used in discussing velocities, to avoid confusion.

So, as a procedure, find, or guess at, a velocity for that reflection in that area, divide the velocity by two, and multiply that by the various reflection times, to get the depths to the formation. On record sections, the time has been corrected to an arbitrary datum plane, which is noted on the labeling of the section, so the depth you get is the depth below that datum. Offshore, the datum plane is normally sea level, or near it. That is, some sections are not corrected to sea level, but left with the start of the section's timing at the average of source and geophone depths in the water. This is preferred by some, who do not like to tamper with the data any more than necessary. It makes only a very small difference, something like 30 or 40 feet, and is uniform through a set of sections, so it can be ignored.

There was a reason for selecting 10,000 ft/sec as an example. It is that 10,000 ft/sec is a good enough guess to use when you don't know the velocity, or just want a rough idea of the depth, or comparative depths. So, for that kind of overall depth, just multiply the time by 5. Notice that 5, rather than 5,000. This works by using the time, not in seconds, but in milliseconds. Instead of 1.231 seconds, think of it as 1,231 milliseconds, and multiply by 5. If there is velocity information handy, you can fine it down, using 4, or 6.2, or whatever, but in general the 5 will at least give an idea of what is going on.

Some approximate velocity information can be used in well known areas. If a reflection can be recognized by familiarity with the general area, then a line shot through a well location can be tied to the well. The time of the reflection and the depth of the formation in the well will give the velocity to that formation at that location, even though there has not been a velocity survey in the well. Just obtain the depth of the formation in the well, from the datum used on the section, and divide by the time, to get the half-velocity. If there are lines shot at other wells in the vicinity, lateral changes of velocity can be determined.

RMS velocities, from the seismic data, are useful, but, especially for fairly deep horizons, may not have the precision needed for conversion to depth. They were described under "Velocity Analysis" in Chapter 5, Data Processing.

The surest and most exact velocity information comes from a velocity survey in a well, combined with a sonic log.

Velocity Survey

A velocity survey in a well is made by firing a seismic source near the surface, and detecting it with a geophone lowered into the well. This gives an exact measurement of the time required for the sound to go to a known depth.

A special velocity survey crew is arranged for. A well logging unit is also made available, so its cable can be used to lower the geophone. The survey is performed as soon as the well is released for it, so as not to run up rig standby charges. If the well is on land, several shot holes may be drilled near the well, several so that if one collapses, it is not necessary to wait while another is drilled. Offshore, an air gun is lowered into the water from the rig floor. An air gun may also be used on land, by placing it in the well's mud pit. A land-type geophone or group of geophones, built into a streamlined unit, is attached to the logging cable, and lowered to the bottom of the hole. Then, it is pulled up in stages, to preselected depths, preferably evenly spaced, say every 500 feet. A spring presses against one side of the hole, holding the probe tight against the other side, so the geophones have good contact with the earth. At each stop the air gun is fired, or a shot in one of the holes, and the time taken by the sound to reach the down-hole geophone is recorded. A paper record is made of each shot, with the arrival displayed on several traces, at different amplitudes, to aid picking of difficult breaks. It may also be recorded on magnetic tape, so it can later be enhanced by processing.

The shots are not right at the well, so the sound travels at a slant, and so the times are corrected to vertical paths by simple trigonometry. However, for deviated wells, the shots can be placed directly above the geophone at the various depths recorded. The paths are then vertical. For deviated wells, the geophones may be gimbal mounted within the probe, to keep them upright.

At each recording position in the well, the seismic time to that depth has been measured, and the velocity is calculated. This provides good velocity and identification information for a seismic line run through the well location. Reflections can be identified just by their times. That is, a seismic reflection will appear as a strong event a little later than the time to that horizon as measured in the well. On the order of 30 to 60 milliseconds is required for the energy to build up to a good peak or trough on the record section, the highest amplitude of the wavelet.

The velocity survey gives data at only isolated points. To make it continuous, it is combined with data from a sonic log, which is continuous, but does not have the overall correctness of the velocity survey. The combination has both qualities.

A sonic log is used for porosity information by geologists and log analysts, but the quantity it actually measures is sound velocity. The sonic log is a continuous measurement of the velocity of sound in the near vicinity of the well. Porous rocks transmit sound more slowly than solid rocks, so the velocities correlate well with porosity.

An instrument containing, in its simplest form, a sound source, or beeper, on one end, and a geophone in the other, is lowered to the bottom of the well. The first energy from the beeper to reach the geophone goes through the rock alongside the well bore, as the velocity of sound in the rock is much faster than in the drilling mud in the well. The instrument is slowly pulled up the hole, with the beeper emitting signals at frequent intervals. The times recorded for the different depths are plotted as a continuous log of velocity.

There may be some cumulative error, though, in the overall velocity picture as assembled from these bits. Mud invasion of the rock near the hole, and washing out, enlargement of the hole diameter, may modify it.

The velocity survey, with its velocities measured by firing shots at the surface, can be used for the velocities overall, and the sonic log corrected to it. The log then becomes a CVL, continuous velocity log, calibrated sonic log, or sonic log with reference marks.

In this combination is the reason for the spacing at even intervals of the geophone positions in the well. At one time, before sonic logs were developed, the shots provided the only data for the survey. So they needed to be taken at formation tops, to yield velocities of formations, rather than lumping parts of formations together. Now, with the shots used only to adjust the sonic log data, they are evenly spaced to distribute the correction uniformly, and the sonic log gives the information on velocity changes at formation contacts.

Synthetic Seismogram

When a velocity survey is made in a well, very detailed velocity information becomes available. Velocities are obtained from a sonic log, calibration of these velocities from check shots, and densities from a gamma ray log. As seismic reflections are a result of velocity and density contrasts, there is enough data to calculate just where in the section there would be seismic reflections, and their amplitudes and polarities. So a "synthetic seismogram" can be made, a theoretical seismic trace, like one that would be expected if a seismic record was made at that location. But the synthetic will show just what contribution was made by each rock layer. Being made from the well data, it is precisely fitted to the well.

The word "seismogram" is an outdated and otherwise rarely used term for a seismic record, that happened to be preserved in this one context.

To make a synthetic seismogram, the velocity and density data, or the velocity data alone (Fig. 7-1), are used to calculate a "reflection coefficient" at each change in velocity and density (Fig. 7-2). The

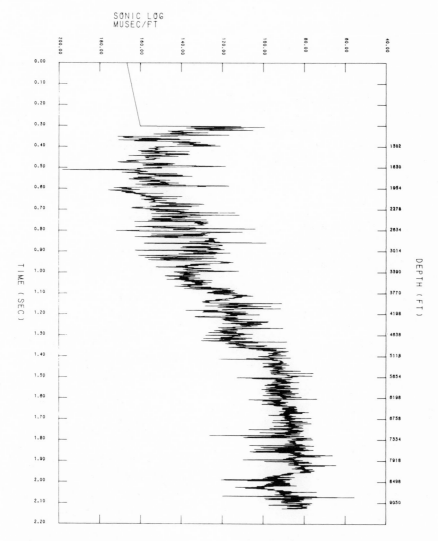

Fig. 7-1 Sonic log at a scale of time   Credit: Western Geophysical Co.

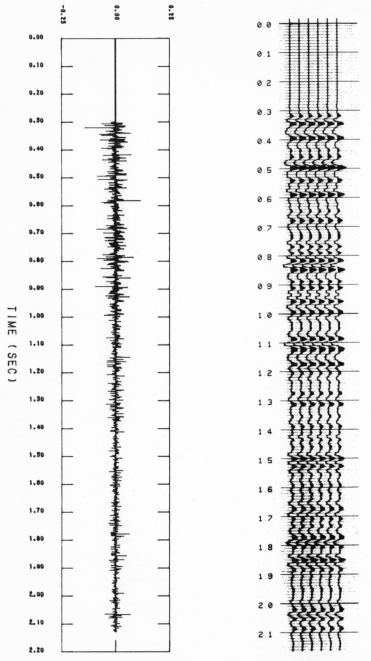

T I M E (S E C)

Fig. 7-2 Reflection spikes from sonic log   Fig. 7-3 Synthetic seismogram

Credit Western Geophysical Co.

reflection coefficient is a mathematical expression of the kind of reflection that will be produced by that interface—amplitude, i.e., how far the trace swings; polarity, which way it swings first, right or left.

The computer program, at an interface, calculates a wavelet of the proper maximum amplitude and the proper initial polarity. Then it does the same for the next interface. But the second interface may be only 5 milliseconds from the first, so the two wavelets overlap. Never mind, the same thing happens to seismic energy in the ground. The two wavelets are convolved, combined into one, more complex, trace, as was described in Convolution, Chapter 5, Data Processing. This goes on for the entire length of the well survey, combining wavelets of varied amplitudes and polarities to make a seismic trace.

Then using the reflection coefficients for energy traveling upward, the amount of sound reflected back down to produce multiples, can be calculated. So multiples can be added to the trace.

The deeper part can be weakened to correspond to the loss of energy in penetrating into the earth, and in spreading out in all directions.

The trace is then displayed at the scale of a record section, so it can be compared directly with sections (Fig. 7-3). The trace is produced four or five times, side by side, to look a little more like a record section.

Now, suppose you have a synthetic seismogram from a well, and a record section that was shot across that same well. Comparing the synthetic with the section, can do several useful things, if they look alike, and sometimes, disappointingly, they don't. Assuming the two are similar, reflections can be identified well. The synthetic is made at the scale, time scale, of the section. But a depth scale can also be plotted on it, and formation tops from the well put on by a geologist familiar with the area. So, the reflection on the synthetic will be expected to be strong at a point about 30 milliseconds later than the interface producing it. And this reflection should match a reflection on the section.

The reflections and multiples were calculated separately for the synthetic, so they can be displayed separately. It is convenient, on a synthetic seismogram, to display a few traces each, of primary reflections only, multiples only, and the combination of the two. The combination will look most like the seismic section. The primary and multiple traces will allow primaries and multiples to be identified as such on the section.

Displays of synthetic seismograms by different companies take different forms, and are made in differing degrees of sophistication. They can range all the way from using sonic data only, no density; and calculating reflections only, no multiples, to the most complete. So do not expect them all to look alike, or to be equally useful. It is best, before ordering a synthetic seismogram, to discuss scales, purposes, special requirements with the company that will make it.

All the above involves straightforward use of synthetic seismograms, at the wells where they are produced. But there are some extra advantages that can be gained from them, that are an aid in stratigraphic exploration.

Pinchouts are difficult to pinpoint on seismic sections, as an apparent pinchout on the section is not where a formation thins down to zero thickness and disappears. Rather, at some place before that, the reflections above and below are so close together that they interfere with each other, so their characters are no longer recognizable. The changed, combined reflection continues, and continues to change in character, but how do you know which character represents the point of pinchout?

Suppose there is a well, with a synthetic seismogram, and the well encountered 433 ft of a formation. This formation is absent in another well, and it would be worthwhile to find exactly where the formation ends. Take the log or logs (sonic and gamma ray) the synthetic was made from. With scissors, cut out the part of the log that shows some portion of the formation, and tape the remaining pieces together. Have a synthetic seismogram made from the spliced log. Cut some more out. Have another synthetic made. Continue until you have cut out all of the formation. Then have the series of synthetics displayed side by side as a record section. This section will show what appearance to expect where the formation is 200 feet thick, 100 feet thick, and where it ends. If no other factors interfere, then this section made of synthetics can be used to determine the various reflection characters on the seismic section.

Similar treatment can be used for reef thickness, salt thickness, and even varying velocities, if you want to modify a bit of the log by moving part of the curve to the right or left.

The actual cutting with scissors is not necessary, but was used to make the point clearly. Processors can perform the log modifications.

Synthetic Sonic Log

A synthetic seismogram is a seismic-like trace produced from the velocity information in a sonic log—and, to a lesser extent, the density information in a density log. Comparing it with seismic traces partly bridges the gap between well data and seismic data.

Another way to compare the two kinds of information is to go the other way around—make something like a sonic log from a seismic trace. Making a "synthetic sonic log" from a trace requries much better velocity information than is usually obtained in seismic work. A sonic log is a log of velocities.

To get this velocity information, something beyond the resolving power of velocities obtained from normal moveout is required.

The amplitude of a seismic reflection is dependent on the velocity contrast of the reflecting interface. If the velocities of the two layers are known, the strength of the reflection can be calculated. So, turning that around, if the reflection amplitude is known, maybe the velocities can be obtained. At least, if the amplitude and one velocity are known, the other velocity can be calculated. A way to have one velocity is to start at the top of the record, where a near-surface velocity can be determined fairly readily. Then use that velocity and the amplitude of a reflection at the base of the layer having that velocity, to determine the velocity under the reflection. Having the second velocity, a third, below the next reflection, can be determined, etc.

For so much to depend on the amplitudes of reflections, those amplitudes must be accurately measured, which requires recording by floating point or binary gain, methods that determine the true amplitude, not adjusted to look strong on a record. Also, for the measurements to show velocity changes in fairly fine detail, the recording should be "broad band", that is, with more frequencies recorded than are normally used for seismic data. In particular, higher frequencies must be included (Fig. 7-4).

Normal moveout velocity data can also be added to the information obtained from the reflection amplitudes, to include all the information possible.

A section can then be made up of traces that look like a series of sonic logs, but considerably smoothed, as even the broad band amplitude data does not have nearly the fine detail of a well log (Fig. 7-5).

As an additional aid to interpretation, the velocities of the points on the traces can be computer contoured. As velocities tend to be distributed laterally, the contours tend to enclose horizontal bands of the section, which then represent various lithologies within the stratigraphic section. In particular, porous zones within reefs, sands, etc. can be interpreted on it. And if the line crosses a well, the sonic log from the well can be smoothed to match the smoothness of the traces,

SEISMIC TIME

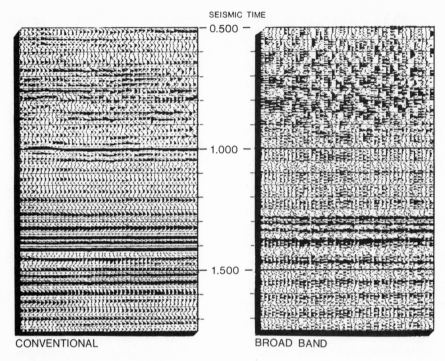

CONVENTIONAL   BROAD BAND

Fig. 7-4 Broad band recording   *Credit: Teknica Inc.*

reduced to the scale of the section, and superimposed on the section as a calibration of the seismic-derived log traces.

As reflections are generally poorer in the deeper part of the section, resolution with depth decreases. This technique is best at reasonably shallow depths, but resolution of strong events below 10,000 ft is common.

Modeling

If the velocity of sound in a rock layer is constant, the path of seismic energy through the layer will be a straight line. If the velocity increases noticeably with depth, at a known rate, the path will be a predictable curve. Then, when the sound enters another layer, its angle will change by refraction, and it will go through that layer in a predictable way.

So, if the thicknesses and velocities of the layers are known, the routes of seismic energy through them can be determined. This is

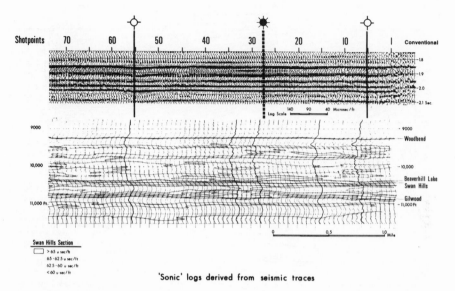

Log Scale 140 90 40 Microsec/ft

Swan Hills Section
☐ > 65 u sec/ft
65 -62.5 u sec/ft
62.5 -60 u sec/ft
< 60 u sec/ft

'Sonic' logs derived from seismic traces

Fig. 7-5 Seislog* synthetic log   *Credit: Teknica Inc.*

called ray tracing (Fig. 7-6). So, if the geologic cross section is known, the seismic section can be calculated. So, like a super synthetic seismogram, an artificial record section can be created.

The value of this is as a check on an interpretation. If a line is shot, the data processed to form a record section, and the section interpreted, the interpretation is a "best guess" as to the real geologic subsurface. But if the geology obtained from the interpretation is made rather complete, with velocities assigned to all the layers, then coefficients of reflection can be calculated and a model made of the seismic section that would be obtained if *this* geology was real, and a line was shot over it. If this new section is not like the original, then something is wrong with some part of the process of interpretation and modeling.

The modeling process is costly and time consuming, so it can not be used regularly, but must be saved for critical points. And the determination and/or assuming of all the velocity data is a lot more interpreting than is necessary in normal work, where the configuration of the beds is the main thing that is needed.

A particularly useful application of modeling is in bright spot analysis. The velocity effect of gas in a formation can be added to the calculations, to see if that aspect of the original section is duplicated, and determine from that an idea of the proportion of gas in the rock.

---

*Seislog is a registered trademark of Teknica Inc.

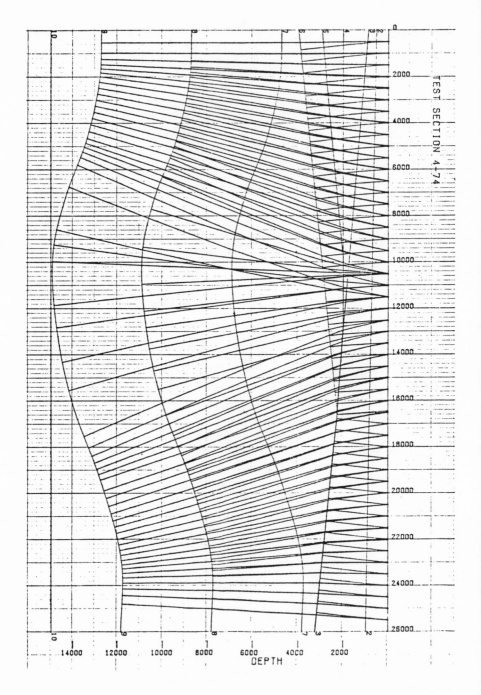

Fig. 7-6 Ray tracing   Credit: Western Geophysical Co.

This is described more fully under the heading "Bright Spot", in the next chapter.

True Amplitude

Seismic data is routinely recorded in floating point form (Chapter 4), retaining a record of the exact amplitudes of the energy. This precision can be used to produce a "true amplitude" section, which has true relative amplitudes from trace to trace. Necessarily, the shallow data is restrained and the deeper is built up to keep them all visible on the section, but the increase is made at a steady rate, so amplitudes are also comparable vertically.

A section so made can be used to compare strengths of reflections, as discussed in Chapter 8 in "Bright Spot."

Dip Migration

A seismic section can be assumed to represent a cross section of the earth. The assumption works best when layers are flat, fairly well when they have gentle dip. With steeper dip, the assumption breaks down, and the reflections are in the wrong places and have the wrong amounts of dip. Dip migration, or just migration, is the process of moving the reflections to their proper places, with their correct amounts of dip.

In a simple example, a trace from a geophone near the shot point, the energy goes almost straight down to the reflecting horizon and back up—if the reflector is flat (Fig. 7-7). If the horizon dips, the energy goes to and from it by the most direct route, which is perpendicular to the reflector. Energy that hits at other angles will go off in other directions (Fig. 7-8).

This perpendicular reflection is the basic idea behind dip migration. All the rest of migration follows from the perpendicular reflection principle. Structure and velocities will cause the sound to follow non-straight paths down to the horizon and back up, but, right at the reflecting surface, the energy path will be perpendicular to it.

So the reflection point is not under the shot point, but offset from it. And the geometry makes it be offset in the updip direction, up the slope of the bed.

That is what happens to the reflected sound. But how does it appear on a record section? The traces on a section are parallel. They all hang straight down from the surface, because they are just measurements of

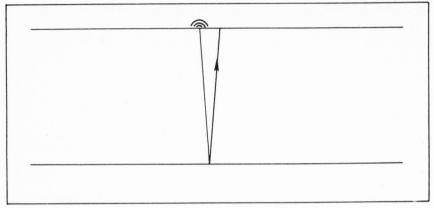

Fig. 7-7 Near-vertical reflection

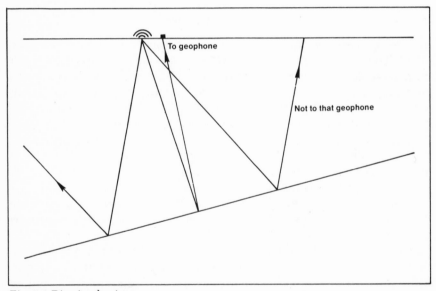

Fig. 7-8 Dipping horizon

the times for a sequence of sounds to strike the geophones. So the time during which sound traveled at a slant, is shown on the section as though the path had been straight down (Fig. 7-9).

Using this perpendicular reflection principle, some subsurface features and how they will look when converted to record section form with vertical traces can be considered. Then some rules can be formed for how the features on the record section will have to change to be migrated back to the correct configurations. For simplicity at this stage, it will be assumed that the velocity of sound is constant all through the geologic section, and that the sections are shot in the direction of dip, so they do not have any reflections from one side or the other of the line.

1. The ray paths—routes the sound travels—are at an angle, and when displayed on a record section as traces, move downward to the vertical arrangement. So migration must move them back updip (Fig. 7-10). Therefore, in migration:

    Reflections move updip.

2. That is just dip in one direction. Now picture an anticline. In the subsurface, the ray paths reach inward to reflect perpendicularly from the horizon. Then, on a record section, they will swing down to the vertical, becoming more spread out, so the feature looks broader. So, migration back to the correct position will narrow the structure.

    In migration:

    Anticlines become narrower.

3. Also in the anticline, the traces swinging down to the record section position show more relief. So migration will reduce the amount of relief (Fig. 7-11). However, this is true only if the anticline is isolated on the record section. But if dips at the flanks become flat, the total relief will not change, as the flat parts are not affected by migration.

    Migrating:

    Anticlines may have less or the same vertical closure.

4. The very crest of the anticline doesn't move. Right at the top, there isn't any dip, so that part won't go anywhere. The energy path there

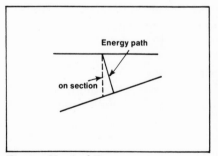

Fig. 7-9 Vertical Trace

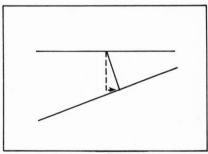

Fig. 7-10 Migration moves data updip

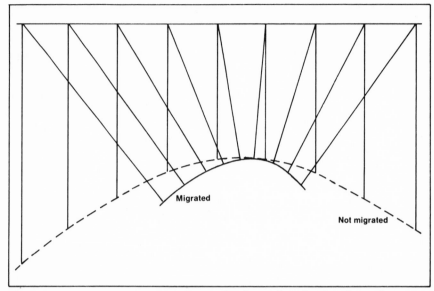

Fig. 7-11 Migration narrows anticline and reduces dip of isolated anticline

is vertical, so doesn't undergo any change in being put in record
section form.

So:

The crest of an anticline does not move.

5. Now in a syncline, the ray paths have to reach out to be reflected
   perpendicularly. So, on a record section, they will make a
   narrower-looking feature. So migration must spread it out again
   (Fig. 7-12).

   When migrated:

   Synclines become broader.

6. And like the anticline's top, the low point of the syncline doesn't
   have any dip, so energy goes straight down to it, and isn't changed
   on a record section, or in being migrated.

   Migrating:

   The low point of a syncline does not move.

7. The amount of relief of a syncline isn't very important, as it's the
   anticlines that have the oil. It would, though, be an inversion of the
   anticline situation.

   Migrated:

   Synclines may have more or the same closure.

8. Synclines can behave in another way on record sections though,

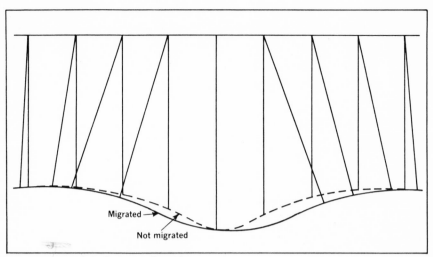

Fig. 7-12 *Syncline becomes broader*

if they are relatively deep in the section or narrow. The deeper, or sharper, ones have ray paths that cross on the way down, with one trace being in a position to receive information from two or even three parts of the syncline. Two crossing lineups of energy, with an apparent anticline perhaps visible beneath them, will be on the section (Fig. 7-13a & b). This is spoken of as a buried focus, as the seismic energy crosses somewhat as light rays when focused by a lens.

Migrated:

Crossing reflections may become a sharp syncline.

9. Now there is a still more involved situation. Where a fault breaks a formation off sharply, or for some other reason there is a point (or edge) in the subsurface, that point returns energy to any source within range, (Fig. 7-14). That is, it behaves as a new source of energy, or looked at another way, as a tiny effectively spherical reflector that is perpendicular to every ray path. So energy is returned to a number of shot points at different distances from it. Then in cross section, with the reflected energy vertically below the shot points, it will become an apparent anticline. It is a very regular one though, like an upright, open umbrella. The form is actually a hyperbola. This point situation is a diffraction, not to be confused with refraction, the bending of energy in going from one velocity to another. Diffraction is generally recognized easily by its regular

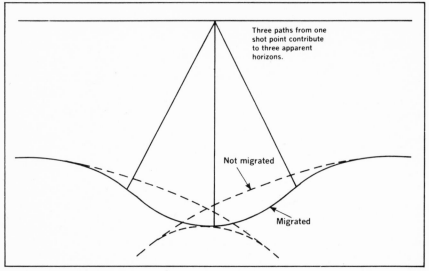

Fig. 7-13a Crossing energy paths

shape. Sometimes only half of it is visible, so the broken-off formation appears to continue in a smooth curve downward, (Fig. 7-15).

Now even though it isn't a normal reflection, a diffraction pattern is created by the seismic traces hanging straight down, so the same process of migration applies, moving the ray paths to their correct positions. So, in migrating the diffraction, it will fall back into the point.

When migrated:

An umbrella shape, diffraction, becomes a point.

10. Also, the diffraction pattern has no dip at its crest, so that part won't be migrated. The crest is also at the isolated point producing the diffraction.

Migrating:

The crest of a diffraction does not move, and is the diffraction point.

Summing up these rules, the changes seen when a seismic section is migrated are that:

1. Reflections move updip.
2. Anticlines become narrower.
3. Anticlines have less or the same vertical closure.
4. The crest of an anticline does not move.
5. Synclines become broader.

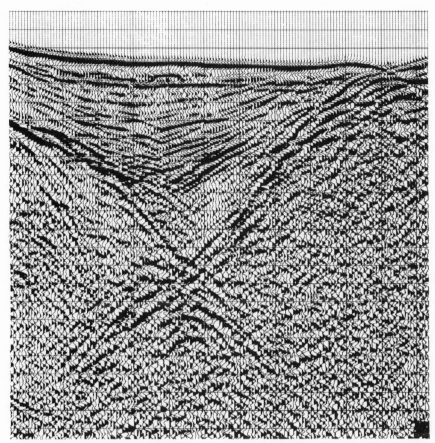

Fig. 7-13b Crossing energy paths   *Credit: Western Geophysical Co.*

6. The low point of a syncline does not move.
7. Synclines have more or the same closure.
8. Crossing reflections may become a sharp syncline.
9. An umbrella shape, diffraction, becomes a point.
10. The crest of a diffraction does not move, and is the diffraction point.

Diffraction Collapse

The hyperbolic diffraction pattern, caused by a point in the subsurface (described in Rule 9 of the preceding topic, Dip Migration), has within it, its own means of correction. The energy that forms the curve, all comes

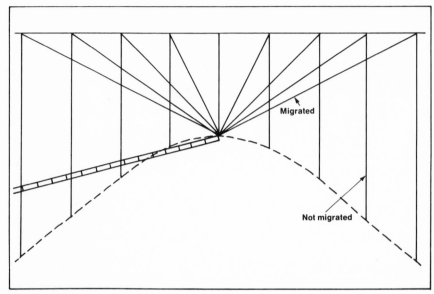

Fig. 7-14 Point in subsurface

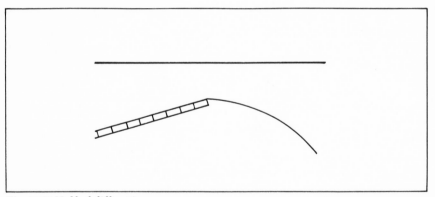

Fig. 7-15 Half of diffraction

from the point at the top of the curve, so to correct it by eyeball, just think of it as being at that point only.

If only part of the curve is visible, a diffraction hyperbola could be calculated for that particular part of the section, using the velocity for that reflection time. This calculated curve would show where the top point of the curve is. The fragment is from that point.

Now the geometry of traces being vertical when they should slant, is the same for reflections as for diffractions. So a reflection can also be migrated by finding the diffraction hyperbola it fits, and moving it to the crest of that hyperbola.

A computer can plot a set of diffraction curves for the appropriate velocities, all the way down a section. This diffraction chart can be on transparent film. Then, overlaying it on a section, you can find where any bit of energy migrates to. Just shift the chart sideways till the energy matches a curve. It belongs at the top of that curve.

A computer migrating system called diffraction collapse, or maximum convexity, uses this principle. It just stacks a lot of traces, along diffraction curves. That is, all energy found on traces along a curve is added together at the top of the curve. This is done all down the section, then the curves are moved over one trace and stacked again, etc. When an event falls on a curve, the top point of that curve has an accumulation of energy. If a curve encounters nothing that fits it, random peaks and troughs tend to cancel, so very little energy is stacked at the top of that curve.

The result is a section with all the events in their migrated positions, a migrated section. If it is well made, and the velocities are nearly correct, it is a much improved section, looking much like the subsurface.

A minor disadvantage of a section migrated by diffraction collapse, is the presence of "noise arcs". These are arcs, open side up, occurring in the deeper part of the section, where they are not many reflections, so random noise creates these artificial shapes. They give the bottom of the section a scalloped look, but are not much problem, as they can be easily recognized and ignored. They don't interfere much with reflections, as where there are strong reflections, the noise arcs don't occur.

Also, diffraction collapse migration produces a section with less of the distinctive characteristics of the reflections, so they are not so recognizable as on unmigrated sections.

Wave Equation Migration

Wave equation migration is a new method of computer migration, that has recently taken over from the diffraction collapse method.

It uses an entirely different approach. From the data received at the geophones, an entire new set of data is calculated—the data as it was just a little time before it arrived at the geophones. In effect, though not usually plotted, it is another seismic section, the one that would be recorded if the geophones were not on top of the ground, buried a little way down. Then another section, in effect a little deeper, is calculated, and so on, all the way down to the reflecting horizons, and on to the bottom of the section.

As each bit of raypath is retraced, things are put in their correct directional relationships. That is, they are migrated. It is somewhat as if you just migrated the shallowest reflection on the section, then used it as a basis for migrating the next one, etc.

The resulting migrated section retains the reflection character, the distinctive reflection appearances, of the original section. This non-degredation of the reflection quality of the section is one of the main advantages of this migration method (Fig. 7-16).

A disadvantage of wave equation migration is that it does not, in its pure form, handle steep dips well. So various modifications have been developed, giving it some of the steep-dip handling characteristics of diffraction collapse migration.

Comments On Migration

Some general comments can be made about migration.

The fact that record sections can be migrated by computer after all other processing is done has an influence on the decision of which magnetic tapes to keep. To save storage, sometimes only the field tapes from an area are retained, and the intermediate and final tapes erased. Now, the final tapes could be migrated without having to re-do the steps leading up to them.

This is only one consideration though, and does not necessarily justify future saving of great numbers of tapes, just in case it might later be useful to apply this or some other technique on the final tapes. It does justify retaining final tapes of some areas with steep dip that you do not now want to spend the money to migrate. The area may become hotter some day, and migration is becoming cheaper.

For many reasons, there will be some doubt as to whether it is worth the time and expense to migrate the dips. In low-relief areas, it is rarely necessary. In intensely folded areas, it is pretty essential. But there are a lot of situations less clear-cut than these two. The best approach to use when in doubt is to have some of the data migrated and see how critically this changes the picture.

Lines that run in the general direction of dip are the ones that most need to be migrated. Lines in very different directions can be misleading after migration.

Computer migration is not dependent on having CDP data. It can be performed as readily on 100% data. The final result will not be as good, but the improvement will be just as great.

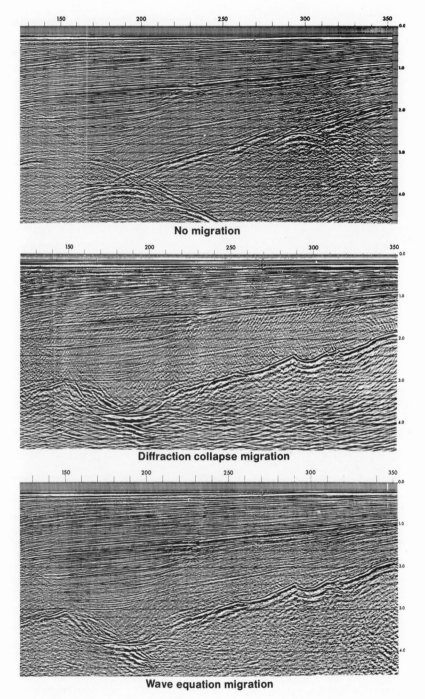

No migration

Diffraction collapse migration

Wave equation migration

Fig. 7-16 Migration of sections   Credit: Seiscom-Delta Inc.

Computer migration of stacked data is customarily performed after stack, for economic reasons (Fig. 7-17). Migration before stack yields better sections (Fig. 7-18), but for 2400% shooting, there are twenty-four times as many traces to migrate. For special problems, it is sometimes worth the price to migrate before stack.

Ties In Migration

A set of unmigrated sections of an area is a representation of the subsurface, but with the distortion caused by all data being shown as directly below the surface points. The distortion behaves in the same way for all lines, and they are compatible with each other. In particular,

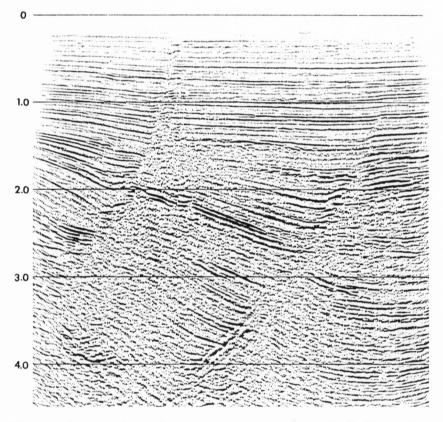

Fig. 7-17 Migration after stack   Credit: Digicon Inc.

they tie at intersections. Both lines at an intersection are distorted by a reflection's being moved from its actual position to a point below the intersection of the lines on the ground. So it is the same reflection on both lines, reflected from the same point in the subsurface.

In migrating, if the reflection on both sections is moved to its true position in three dimensions, then the migrated sections will also tie.

However, normal migrated sections are migrated in two dimensions only. This migration does not treat intersecting lines alike. If one line extends down dip, the migration of reflections on it will be correct, putting them in their true positions. The intersecting line, if the lines intersect at a right angle, will run along strike, perpendicular to dip. It will exhibit no visible dip, and migration will not change it. So, at the

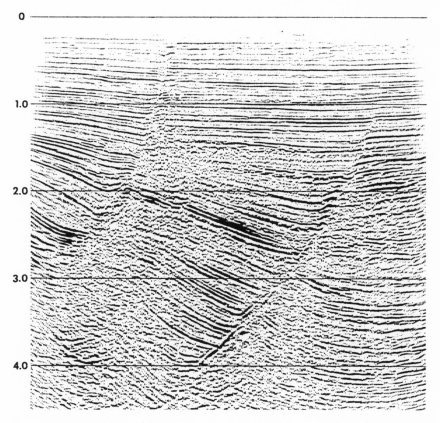

Fig. 7-18 Migration before stack   Credit: Digicon Inc.

intersection, the reflection on one line will be moved to its true position, while the same reflection on the other line, will not be moved at all.

The lines will not tie.

So, migrating all sections to get the "best" seismic data, will not even let you go from one section to another. What is the best thing to do?

To handle the data, the unmigrated sections should be used to provide line ties. The sections that run generally down dip, need most to be migrated, and yield the most benefit from it. In some cases, lines more or less along strike may benefit from migration, in sorting out features. The choice of which lines to migrate is a matter of judgement. It may be advisable to migrate all lines in some areas. But these two-dimensionally migrated sections cannot be used to tie intersections. So the reflections can be picked on the unmigrated sections, with guidance from the migrated sections. Details of faulting, etc., will be determined on the migrated sections, and can be used in working the unmigrated ones. Data from unmigrated sections can be plotted on a map, and if necessary can then be migrated in three dimensions. Or, some features of the migrated sections can be added to a map.

The key, migrated, sections will play a big part in drilling decisions, and may be picked and displayed to show the kind of prospect encountered.

Even in an area without steep dips, there may be some reason to migrate the sections. Migration may show small faults more clearly, eliminate confusing diffraction patterns at reef edges, etc. But with little dip in the area, the tie problem will not be great. It may be that in a situation like this, sections, both migrated and not migrated, could be put together at intersections, and would tie reliably.

Three Dimensional Migration

All the migrating described so far has been in two dimensions—along a record section. This is fine if the section happens to be aligned along dip—perpendicular to strike. But there are a lot of reasons why the line may run in some other direction:

Terrain may dictate the direction of the line,

The line may have been laid out between two wells,

It may be a tie line between other lines,

The direction of dip isn't known before shooting—except in a regional sense,

The direction of dip on one horizon may not be the same as that of another horizon, so a single downdip line for all horizons isn't even possible.

If the line isn't a dip line, then migration along the line isn't correct. Updip—the direction things migrate—isn't in the plane of the section.

So, what is to be done about this problem? For one thing if the line is fairly nearly a dip line, or if there isn't much dip, the problem can safely be ignored. When it can't though, the data can be migrated in three dimensions. There are several ways to do this. A simple, though laborious, manual one follows.

The data for one horizon are plotted and contoured on a map, in regular unmigrated form. A line extending down dip is drawn on the map, perpendicular to the contours, anywhere that contours are nearly parallel for some distance. Contour values at intersections of contours with this line are plotted by hand on a cross section. Some migrating method is used to migrate this section. The computer methods described don't apply, as they need a record section as a starting point, but overlaying a diffraction chart will work. From the migrated horizon on the cross section, data are plotted on a new map. The process is repeated for other dip lines. When enough migrated points are plotted on the new map, it can be contoured. It is then a map with data in true migrated position.

The process has to be repeated for other horizons. Note that the dip lines of one horizon are not necessarily dip lines on another, so a whole new set of cross sections may need to be plotted for each horizon.

This three-dimensional migration points up certain aspects of seismic lines. One is the curved line. A seismic line is laid out as a straight line if at all possible, so the data will stack properly. However, there are occasions when a line must curve or not be shot at all. Examples are marine-type shooting in a river, which must follow the river's course, and land shooting in a narrow and winding valley. These curved lines constitute a very special situation when there is also steep dip in the area.

Consider the influence direction of shooting has, first, on straight lines. If a line is shot along dip, the correct dip will be represented on the record section (Fig. 7-19). If, however, the line is shot along strike, no dip at all will be recorded (Fig. 7-20). The horizon will look perfectly flat on the record section. Anywhere between the two, the section will show some dip, but less than the actual amount (Fig. 7-21).

If the line curves, wandering up regional dip and then down, an apparent structure will show up on the record section (Fig. 7-22). It may look geologically reasonable, with less dip in the shallow beds than the

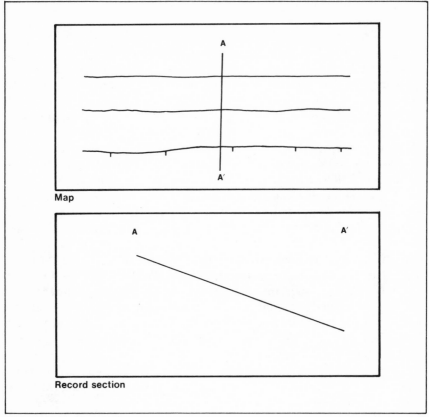

Fig. 7-19 Dip line

deep. So, if the bend is not noticed, and the line is migrated as a two dimensional line, assuming it is about straight, it can be extremely misleading.

In three dimensions, of course, the data would migrate in a true updip direction. If the line had been shot as two straight lines, with the bend skipped, the data on each line would move in the updip direction for the lines. But, if the curved line is migrated in two dimensions, say by diffraction collapse, what happens? The segments are migrated around the bend. Now, even regional dip is turned around, and on the recorded section a syncline appears.

That is a very extreme case, with its hairpin turn, and would probably never happen. It illustrates the point, though, and less sharp bends will

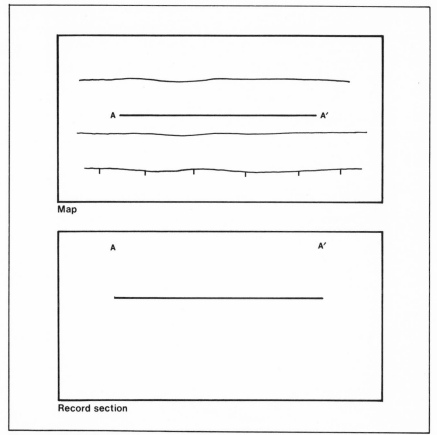

Fig. 7-20 Strike line

produce some of the same type of confusion. They will make dips do confusing and incorrect things. The moral of this is just that record sections can't safely be migrated without thinking of special factors that may influence them.

There are several computer programs for migrating in three dimensions. The data may be migrated, not in the form of a map of a horizon, but as record sections. That is, the energy on a close grid of seismic lines is three-dimensionally migrated. Then the data is moved, not only along a line, but sideways also, so some of it may even wind up on another line. Then sections can be produced with all data in its true three dimensional positions. Sections can also be produced from intermediate locations

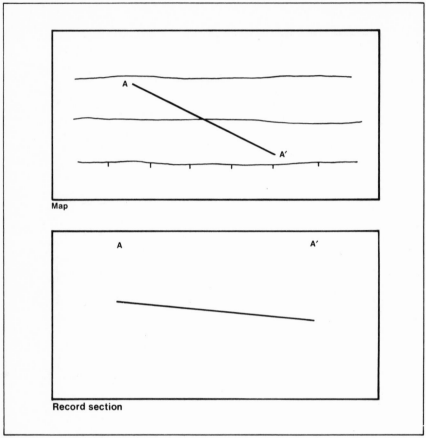

Fig. 7-21 Intermediate line

between lines, by picking off the data that migrate there. So sections can be made in any direction also.

3-D Shooting

Seismic data is almost always obtained in straight lines, so the information is essentially two dimensional, the plane below the line, although the subsurface is three dimensional. This presents problems when data is migrated along a line that is not in a dip direction, when

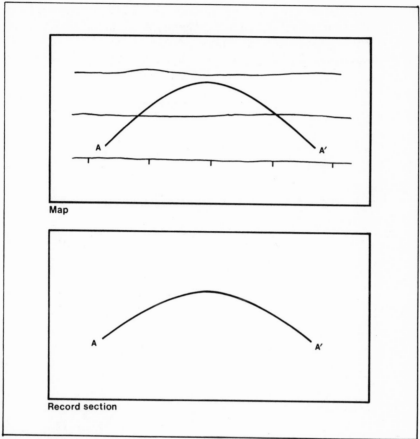

Fig. 7-22 Curving line

there are reflections from the side, when small features are found on a line (where is the highest part of the feature?), when faults are located (which direction does the fault trace go?).

There are several ways to obtain sufficiently dense three dimensional data in the field. All take advantage of the fact that subsurface data is obtained from the midpoints between source and receiver, arranging shots and geophones so that the midpoints are not in the usual straight line, but spread out over an area.

If two or three lines of geophones are laid out, parallel to each other, and a line of shots, also parallel to them, the depth points will be in parallel lines midway between lines of geophones and line of shots (Fig. 7-23). Or there can be several lines of shots and one of geophones.

A line can be shot with geophones, shots, or both in a crooked line,

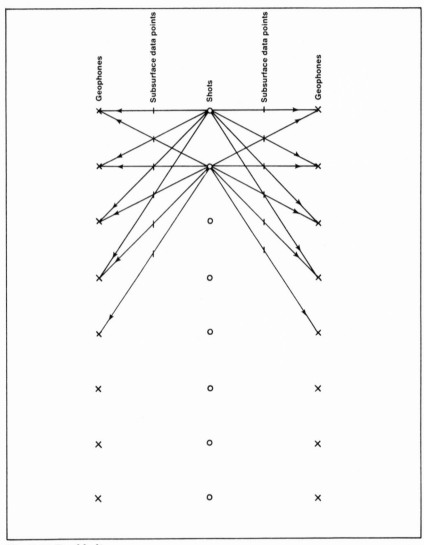

Fig. 7-23 Double line

perhaps following a trail or river. A midpoint between a shot and a geophone around a bend will not be on the trail. There will be a scattering of points near the trail (Fig. 7-24).

A line of shots and a line of geophones can be perpendicular to each other. In this situation the midpoints form a set of lines parallel to, and half as long as, the source or receiver line (Fig. 7-25).

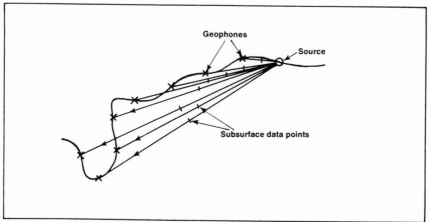

Fig. 7-24 Crooked line

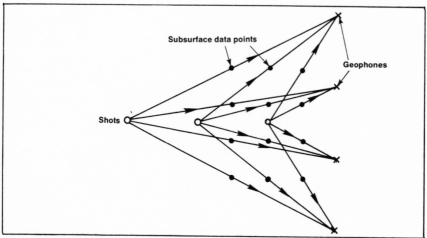

Fig. 7-25 Perpendicular shot and geophone lines

A geophone cable can be laid out completely around an area, and shots fired along the cable, also around the area. The midpoints will be mostly or all within the area. The area can be a rectangle or an irregular shape (Fig. 7-26).

In a special offshore application, the feathering of the cable, that is, the bend of the trailing end caused by currents in the sea, can be utilized. If the feathering is great enough the midpoints for the end geophones are offset to the side from the midpoints for those near the boat. So as the boat advances, there is, not a line, but a swathe, of midpoints (Fig. 7-27).

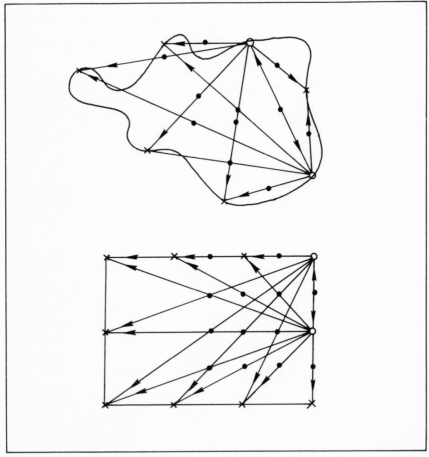

Fig. 7-26 Enclosed areas

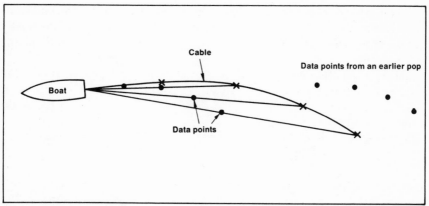

Fig. 7-27 Feathered cable

The exact offset of the cable can be determined by magnetic compass devices that can be installed in the cable, and connected to wiring to transmit information to instruments on the boat. With known length of cable and direction at several points, geophone positions can be determined.

Fig. 7-28 Offshore Oregon-wave equation migration

Credit: Western Geophysical Co.

8

Interpretation

After record sections are produced by the data processors, the sections come to the purpose they were made for. The shooting and processing are done only to obtain seismic sections to be interpreted. And the interpretation, of course, is in turn made as a means of finding places to drill for oil or gas, etc.

Seismic interpretation is the process of determining from record sections information about the subsurface of the earth. It may locate new prospects for drilling, or may guide development of an already discovered field.

The most common activity in seismic section interpretation is mapping a horizon. A horizon is a reflection that appears on sections over some geographical extent. The reflection is identified as representing a certain geological formation, or only as being a reliable reflection. It is "picked", marked, at shot points along the sections over the area. The picks are timed, i.e., reflection times are read. These times can then be plotted on a map and contoured, to show locally high places, or other features that may be prospects for drilling for oil.

There are a number of special problems in interpretation—identifying the reflections, staying on the same reflections throughout an area, presenting the results, etc.

Identification

When a seismic line is shot, if record quality is good, there are a number of reflections on the resulting record section. Similarly, when a well is drilled, a number of formations are encountered. Now if the line and the well are near each other, at least some of the reflections are from some of those formations. But how do you know which is from which? The reflections are measured in seconds and the formations in feet.

If you have an absolutely perfect knowledge of the velocity of sound at that location, you can calculate the relationship and determine exactly how deep a reflection is, or how much time is required for sound to reach a formation and return.

The nearest approach to this perfect velocity knowledge is when the line is shot across the well, and a velocity survey is made in the well. Then, multiplying half the reflection time by velocity gives the depth to the reflector. Or, dividing the depth by the velocity, and doubling, tells what reflection time would represent that formation top.

If there is a velocity survey in one well and a seismic line does not go through that well, but does cross another well in the area, then a correlation from one well to the other can provide a sort of velocity information at the second well, and it can be related to the seismic results.

One of the most common ways of identifying reflections is to compare the record section with another section on which identifications have been made. This breaks down as the lines get farther apart.

The synthetic seismogram is frequently used to identify reflections. It is made from a sonic log, and can be correlated directly with a record section.

Velocities obtained from the seismic data are sometimes good enough to identify reflections in a very general way. In some new areas this may be the only means of identification available.

Also in remote areas, projecting surface geologic data downward may be the only connection between seismic reflections and geologic depths.

In general, in seismic work, a geophysicist can not put his finger on a reflection and say, "There is the So-and-So Formation." This eyeball recognition often works in familiar areas, where much experience has been gained, and a form of it operates in areas known to have only one or two good reflectors. If you expect just two reflections and see them both, then it's easy to tell which is which.

Continuity

Once a horizon has been selected for picking, whether identified or not, there must be some means of picking the same horizon throughout the area, or at least through some portion of the area.

If the record quality is good, and there aren't any complicated situations along the line, a colored pencil mark can be drawn on the reflection from one end of the section to the other.

A comment on the pencils. The introduction of erasable colored pencils is one of the real advances in geophysics. Before they were available, an interpreter had to just use black pencils as long as he wasn't sure of the continuity. And black pencil doesn't show up well on a blue line print of a record section. And especially it doesn't show up on a black line print. Then, after being pretty sure, colored lines could be drawn. They showed up fine, but as soon as they were put on, someone would likely come along with another section that tied this one, and called for changes in the interpretation. The especially erasable colored pencils will allow the interpretation to be made in red or whatever, and changed as often as needed. You can tell if they're erasable by seeing whether the manufacturer provided erasers on them. But don't use those erasers on record sections. They're too abrasive. Use soft erasers, either plastic or gum.

The interpretation problem comes in when the record quality is not so good, or when there is faulting or some other complicating factor. So you have a starting place, the well location or other point at which the reflection was identified. As long as it is apparent that the same energy is continuing without a break, it can be picked. Then if poor records interfere, making a break in continuity, you can look above and below it for better reflections. If there are some, the poor one probably conforms with them, so can be drawn parallel to them, and picked up again on the better records, on a reflection that the drawn line meets or nearly meets (Fig. 8-1).

Occasionally an area will be encountered in which no continuous horizons can be picked at all. This may be because record quality is so poor, or perhaps because the subsurface itself contains no continuous layers, but just lenses of sand and shale, as in parts of the Gulf of Mexico.

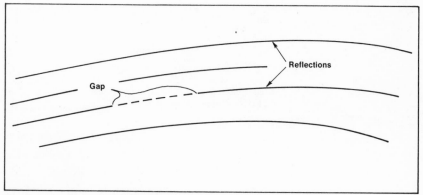

Fig. 8-1 Parallel other reflections

In either of these cases, a line can be drawn on the record section paralleling the short nearby reflections. This line is a phantom horizon (Fig. 8-2).

Also, the part of the reflection already picked can be extended with a straightedge, continuing the same dip, and reflections on the other side of the poor zone extended back toward it. A good meet of these extensions indicates that you may be on the same horizon (Fig. 8-3).

Position on the section is useful in determining which is the same reflection after a gap. Naturally, if you were picking the first strong reflection on the section, then after the gap, the first strong reflection is a candidate for being the same one. And if the horizons are not dipping much in the vicinity, then one at about the same time or depth is likely to be the same one.

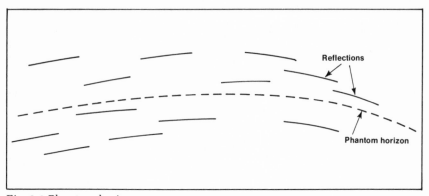

Fig. 8-2 Phantom horizon

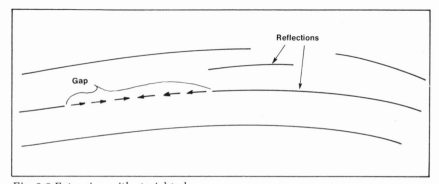

Fig. 8-3 Extension with straightedge

This leads into "character." Reflection character refers to the shape of the wiggles that make up a reflection. It may have come from the word characteristics, which makes more sense in this context, but the word character is ingrained in the language and can't be pried out.

A seismic trace is made up of just sine wave type wiggles back and forth across a central position, but the wiggles can vary in size, and be superimposed on each other to form more complicated wiggles, so there is a good deal of variety in the shape a band of energy can take. A "band of energy" means several peaks and troughs in a sequence. A reflection can be recognized by having one large peak followed by a small one, two large and one small, or something (Fig. 8-4). A broad peak with a dip in its top is called a saddle. Two peaks, or troughs, together are a doublet, or doubleton.

Almost any reflection will have some combination of two or three of these features that persist over some distance on a record section. The combination is the character of the reflection. A character often persists over many miles. An example is the two-peaks character of the Wabamun, recognizable all over the Rainbow-Zama area in Alberta, Canada. In other areas, a recognizable character may carry for only a short distance.

Character permits correlation of traces, just as well logs are correlated, so reflections can be followed across faults, poor records, gaps in con-

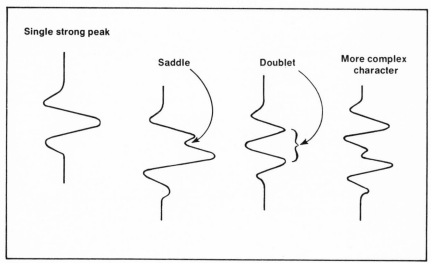

Fig. 8-4 Various reflection characters

trol. The character on seismic records is not as definitely recognizable as are features on well logs, and, of course, not as detailed and measurable down to the foot, either.

In most record section interpretation, except where there are no problems, all of these continuity techniques are used to some extent. In combination, and used by a practiced interpreter, they constitute an effective means of determining horizons.

Loops

Seismic lines are often arranged in loops. That is, they cross each other so a reflection can be followed from one line to another, and on around to its starting point. On a regular grid of lines, the loops will be rectangles. With other programs, they may be irregular shapes with three or more sides. The idea of a loop is, as in transit surveying, to see if you are still on the same level when you get back to the starting point.

Loop tieing is a good check on the picking. If a loop ties, it doesn't necessarily mean it is picked correctly—two errors could cancel each other. But if the loop doesn't tie, the picking has to be wrong somewhere on the line.

When a loop doesn't tie, the first step in correcting it may be to go back to a point where a difficult decision was made, and try it the other way. Another step is to check around the loop for just plain errors. On occasion, there is nothing better that can be done than to find the poorest part of the loop, and assume that the error is there. Then, just change the interpretation there, to make the loop tie. This isn't good, but you can't just wait for a miracle.

In a large area in the Gulf of Mexico, a grid, 2 by 3 miles or something, was being shot. Two factors made good ties difficult. There were a lot of faults in the area, but they did not show up seismically. And the subsurface tended to be made up not so much of continuous beds, as of a mass of lenses of sand and shale. A careful interpretation was being made, but we also needed some idea of the structure quickly. So I instituted my "sloppy map". To make it, I interpreted sections, mistied loops by 100, 300, 500 feet, then just adjusted the lines to force the loops to tie. The adjustment was made with no consideration for quality of data—just take 100 feet of dip out of this line, and 250 out of that one, because it will also help the adjacent loop to tie.

The map wasn't technically correct, but it was continually ready when needed. And, after the more careful mapping was done, the similarity between the two forms of mapping was pretty good. Certainly the salt domes were in the same place on both maps. The point of this story is that in some situations loop tieing alone can produce usable data.

For some situations, loops aren't necessary. A line between two wells, that ties the well data, doesn't usually need further checking. A long line, shot only for regional information, doesn't need to be so correct. And sometimes only an indication of whether there is a higher position near a well is needed.

If several lines are shot in the direction of dip, as in the steeply dipping areas along mountain fronts, they are often left disconnected, or are tied together by only one line joining them. In this case, a second tie line to form loops may or may not help. If there are faults with large throw, the faults may make ties unreliable, so the loops don't produce any increase in confidence in the interpretation.

Formations And Reflections

Seismic record sections are interpreted in terms of geological formations, and the reflections identified as coming from certain formation tops. This is a simplification that usually works well enough for finding oil, but sometimes breaks down, and then seems very puzzling.

Reflection takes place at velocity interfaces. The contact between two formations seems like it ought to be such an interface. Usually it is, as when the two are composed of shale and carbonate. But if the two have the same or nearly the same velocity, then there won't be much reflection there.

So if two formations have about the same velocity at their contact, but one has a lithologic change that varies laterally, say with the lower part sandy and that sandy part thickening in some direction, then the top of the sandy part may be the reflector, rather than the formation contact (Fig. 8-5).

Thickness is another factor. If there is a velocity difference at the top of a layer and another at the bottom of it, then each surface will reflect energy. If they aren't far apart, the reflections from them may come to the surface one after the other, say as two wavelets that do not overlap. If the layer is thinning, then the wavelets will get closer together. At some point, a peak of one and a trough of the other will coincide, so the total energy is less, as the two tend to cancel. When the distances are such that the first peak of one and the second peak of the other coincide, there is reinforcement. So reflection character, the combination of these individual reflection elements, changes with the thickness of the bed (Fig. 8-6). And the character change can appear as a falsely dipping horizon.

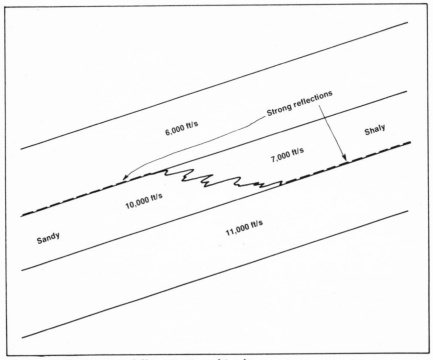

Fig. 8-5 Reflection may follow stratigraphic change

As the bed gets very thin, only character change will indicate the pinching out of the bed. Some distance before the actual zero edge is reached, the reflections will no longer be separately recognizable, so the pinch-out will seem to occur before it actually does. Synthetic seismograms can help resolve this problem, as was described under Synthetic Seismogram in Chapter 7, Data Handling.

In seismic mapping, it is desirable to map a single layer, like the top of a potential producing zone. But that interface isn't alone. There are many little interfaces above and below it. And each of these reflects some energy. So the reflection seen is a composite of them all. If the mapping horizon is a good strong interface with a large velocity difference, it will dominate the others. But even then, the shape of the reflection is affected by elements from different interfaces. So its character remains the same only if the various interfaces maintain their distances apart and their velocity contrasts. Changes in these quantities then, explain why reflection character is often fairly inconstant, and in some areas totally unreli-

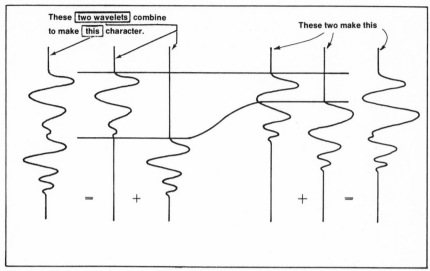

Fig. 8-6 *Character changes with bed thickness*

able. It also explains why reflection character will sometimes give clues to lithologic changes.

With this composite nature of reflections, and the normal shape of the wavelets, with their greatest amplitudes some time after their origins, it is unlikely that the very start, onset, of a reflection can be picked. Instead a later part, after the strength has built up, is recognized and picked. The delay varies, but is about 30 milliseconds. So a reflection that is picked to represent the top of a formation may be a direct result of that top, but occurs some 30 milliseconds, or about 150 feet, deeper. This discrepancy is usually taken care of, whether consciously or not, in selecting a velocity to make the pick fit the depth of the formation. Or if a velocity from another source is used, then adding or subtracting a constant amount to the reflection time will make the data fit. This is not improper. It is a simple adjustment for the imperfection of the data. As long as the adjustment is constant or fits some formula, it is a perfectly reasonable empirical approach.

Also, sometimes it may be advisable to interpret one formation by mapping another. This happens inadvertently sometimes anyway, when a not quite correct identification and a slightly off velocity make a reflection seem to tie wells. An advantage of seismic data is that the sections show pretty well whether the horizons are generally parallel. If they are, then one can be used about as well as another for information about a formation. So, if a nearby reflection is more reliable than the one on the exact horizon, better information about the desired level may be obtained from the better reflection.

So, for a number of reasons, the reflection followed may not be from the formation on which information is desired. Sometimes this is misleading and puzzling. At other times it is valid and useful.

Faults

Faulting is one of the more important parts of a structural situation. But, seismically, faults are very difficult to map correctly.

Under good conditions, a fault can show up clearly on a seismic section. Reflecting horizons have offsets in them, and these breaks on the various horizons follow a slanting path on the section. This path represents the fault plane, as it intersects the seismic line. Faults are clearest on migrated sections (Figs. 8-7a & b).

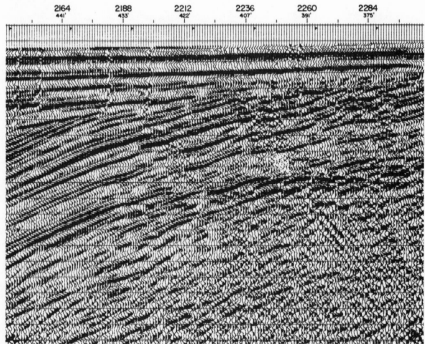

Fig. 8-7a Migration of faults–unmigrated   Credit: Western Geophysical Co.

This is the ideal situation, but even then, there are some problems, concerning the angle of the fault and its direction. And, when faults are more subtle, or records are poorer, the faults are hard to find on the section, and throw is difficult to determine.

There are several kinds of less direct evidence for a fault, when it isn't clearly visible.

A loop may not tie, because a fault was not seen. So it can be sought all around the loop.

A jump in correlation may occur but without a break in continuity. That is, the horizon may seem to be continuous, but two places on the line may appear similar only if that continuity is not followed (Fig. 8-8). This can be pretty subtle. In some cases the correlation is correct, in others the continuity. So it is just a matter of working with the records in the area long enough to become so familiar with them, that you can come to a conclusion with some confidence.

A change in the dip of a horizon, or a number of horizons, *may* indicate faulting. As an example, suppose on part of the line everything

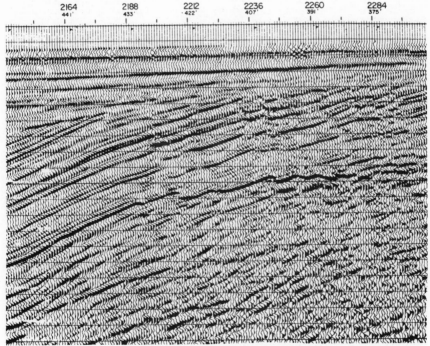

Fig. 8-7b *Migration of faults—migrated*   Credit: *Western Geophysical Co.*

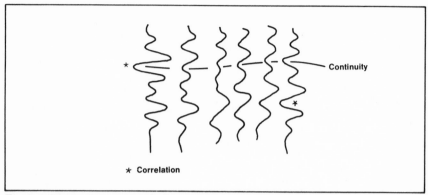

Fig. 8-8 Continuity vs. correlation

is flat, then at some point on each horizon, there is a change to some dip, and the points of change on the different horizons are not one above the other. The hooker here is, of course, that the section may just be indicating the true situation, a change of dip, with no fault. A better example is for the dip to change, then return to its original attitude (Fig. 8-9). Again, the section may just be telling the truth. My own preference, particularly when record quality is good, is to consider that these situations indicate dip only, not faulting, the formations having bent, but not broken.

Any non-vertical break in the appearance of the section—in continuity, character, anything—may be caused by a fault. Non-vertical, because a vertical break may only be a result of a change in the near surface formations, or a processing problem, or could even be a place where two segments of record section film, with slightly different developing, have been spliced together. There can of course be vertical faults, but these problems should be considered before interpreting them.

Local distortion of bedding, in the form of drag against a fault, may be apparent on the section.

Diffractions are a good indication of faulting. Where a formation is broken off by a fault, there is likely to be a diffraction on the section. Usually only half the arc of the diffraction is apparent, the part beyond the end of the formation. So the rock seems not to end, but to curve downward. If the diffraction cuts across other reflections, then it may be apparent, from the superimposed look, that it is a diffraction. At a fault, a number of formations are broken, so there can, in good circumstances, be a slanted alignment of diffractions (Fig. 8-10). Migration, in eliminating

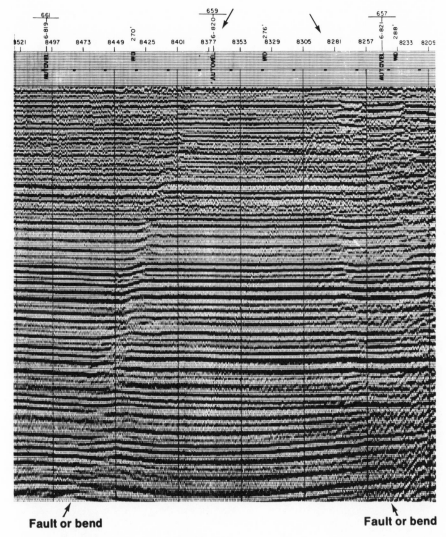

Fault or bend   **Fault or bend**

Fig. 8-9 Possible faults (detail of Fig. 12-1)   *Credit: Teledyne Exploration*

the diffractions, will make the fault more clearly visible, but the diffractions themselves can be seen only on the unmigrated section. So both types are useful in determining faults.

Non-seismic evidence can also help in fault interpretation. If a well cuts a fault, and the well is on the seismic line, then it is saying that there

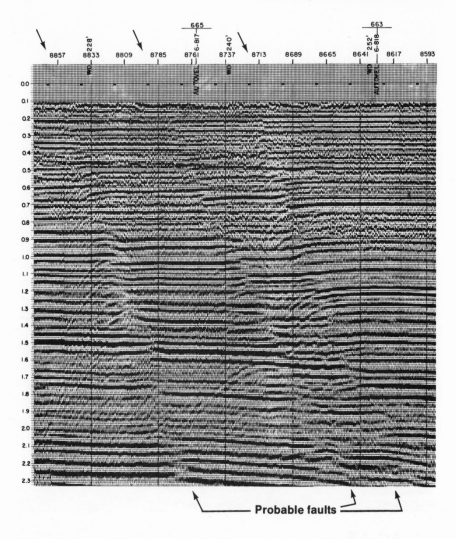

Fig. 8-10 Aligned diffractions (detail of Fig. 12-1) Credit: Teledyne Exploration

is a fault at that point. If it can't be seen on the section, then there is a problem, but usually the well logs and section can be worked out together, and a joint solution arrived at. Either may cause the other to be reinterpreted somewhat. A fault in a well that is not on the line, but only nearby, is less direct evidence, and requires more flexible treatment.

Surface faulting may also be tied in with seismic sections. In some areas it is advantageous to plot surface faulting and other surface geology along the top edge of the section.

A caution is in order in seismic interpretation of faults. In an area known to have faults, there is a strong temptation to think you see faults, on little or no evidence. This is a situation in which it is much better to acknowledge to yourself that you don't know, and so indicate on maps, say by not mapping any faults, and adding a note to the effect that there are probably many faults, but that they cannot be found on the sections. A good guess, on some reasonable basis, is warranted of course, but watch out that you don't kid yourself into thinking you see things that aren't real.

Angle Of Faulting

Where a seismic line crosses a fault, and the fault shows clearly on the section, as a slanting line of broken or definitely offset reflections, and if the seismic line crosses the fault at a right angle, then the angle of the fault seen on the section will represent the true angle of faulting. It will represent it in terms of seismic time as the vertical dimension, and distance as the horizontal. To determine the angle of the fault, use whatever velocity data is available to calculate the depths of two or more points on the fault. Then, using the locations of those points, obtained from the shot point numbers at the top of the section, plot the points on another piece of paper at a one to one scale.

If the seismic line is the one to discover that fault, though, it is very unlikely to cross it perpendicularly. If it is a detail line, shot after the fault was found and mapped, then it could be planned to pretty well cross it at a right angle.

In any other case, the line will cross the fault at less than a right angle, more obliquely. The fault will appear on the section to be less steep. The ultimate situation is when the line runs parallel to the fault, so the fault plane would appear horizontal, if it could be seen at all. So the general case is that an apparent fault plane on a seismic section, after plotting at 1:1 scale, is less steep than the fault itself. Notice that this is the same as the apparent dip situation that was described in "Three Dimensional Migration", Chapter 7, Data Handling. Both problems can be resolved by plotting on a map, and, if necessary, migrating in three dimensions. However, such a step is hardly ever taken for fault interpretation, as the fault can normally be analyzed by eyeball well enough for exploration purposes.

Mapping Faults

The mapping of faults is probably the most severe problem in seismic exploration at present.

A map of a drilled-up salt dome oil field may be largely a criss-cross pattern of faults. A cross section of the same field can also be a similar pattern. These patterns are derived from studies of well logs, production histories, etc., and are themselves fallible, but represent the best picture that can be obtained with much data and much effort by qualified specialists. The point here is that the field truly is cut up by faulting into a hodgepodge three dimensional pattern of blocks.

Then, if a seismic survey is run over a similar prospect, you can be almost certain that a similar complex of faulting exists there. From that it is normal to assume that the faulting should be mapped.

Seismic fault evidence occurs on a seismic section, which is two dimensional. If it shows clearly, the fault trace is a line of broken beds slanting across the section. In a highly-faulted area, there should be a number of these fault traces on a section. Then, on the next section, say a nearby parallel line, there should also be a set of faults. Now comes the interpretation problem. Which fault on the first line connects with which one on the second line? Occasionally, a fault may have a recognizable character—the biggest fault in the area, say—but a common situation is that there is no way to identify a certain fault from line to line.

Something has to be done, to make a map. It won't do to just indicate where faults cross a line. If the interpreter does that, then he hasn't finished interpreting, and someone else has to complete the interpretation. A guess has to be made. And it is usually guided by some idea of the local geology. Then, when the map is complete, there is an interpreted pattern of faults. If the interpreter has been conscientious about portraying the intensely faulted situation he is sure exists, then he will have main faults, with others branching off from them, faults meeting each other, etc. On a drafted map, these faults, some of them guessed-in, all look quite sure and definite. So, wells are planned on the basis of the interpretation. Maybe two faults meet updip, isolating a nice wedge of sediments. "Let's drill here, in the updip corner." "How many acre-feet are there in the prospective field?" "What are the recoverable reserves?" What hogwash! It may be that those faults meet because of a guess. They had to be drawn in some directions, so they have to meet somewhere.

Of course, there are situations in which there is good evidence for the directions of faults, but the general problem is that the faults are visible on the sections, clearly or not so clearly, but there isn't any regularly reliable way to map them.

Seismic data that yields some kind of three dimensional information is the best hope for solving the problem. Several systems for obtaining three dimensional data have been or are being developed. They were discussed in "3-D Shooting" in Chapter 7, Data Handling.

Mapping

Seismic data are in most cases, best displayed in map form. Mapping is a part of the interpretation of the data.

Reflection times are obtained from the record sections. That is, a horizon is marked on the sections, and, at each shot point, or every fifth or so shot point, the reflection time for the horizon is determined. This reflection time is plotted at that shot point's location on the map. Or the time may be first converted to depth, and the depth plotted.

The plotted times, or depths, are contoured. The contours are lines of equal time or depth drawn wandering around the map as dictated by the data. Then, in the simplest situation, the anticlinal trap, oil may be expected at locally high points of the mapping. That is, where the high contours are "closed", go completely around some space, the high area they surround may be a good location for a well.

Other situations involving faulting, pinchouts, reefs, are more complex, but are also displayed and evaluated on seismic maps.

The seismic map is usually the final product of seismic exploration, the one on which the entire operation stands or falls. Selected record sections may be displayed with the map, but the map itself is normally the primary basis for decisions about where, and whether, to drill for oil.

The process of contouring a seismic map is a matter of many choices, the contours being, except right where they cross lines of data, guesses at what would be found if there were shot points everywhere on the map. Thus the shapes of contours, particularly those farther from shot points, should not be taken too literally. The main function of contouring is just to make the numerical data more visual, so the high and low areas may be grasped quickly, without peering at individual numbers. Looking at a seismic map then, should involve, not just seeing what structure the contours show, but also an awareness that contours near data mean more than farther ones. The farther contours had to be drawn some way to avoid leaving holes in the mapping. A refusal to contour away from data would destroy the pictorial aspect of the contouring.

Also, the more "guessed-in" contours point out leads that should perhaps be followed up by more lines of shooting.

Contouring—Geological And Geophysical

Contouring is contouring, and anyone who can contour one type of data can contour another.

That seems apparent, but isn't quite true. In geological contouring, the regional picture and expected shapes of features of the area must influence contouring, if it is to make sense.

Seismic data in a similar way, have their own characteristics that must be taken into account to produce a sensible map.

First, seismic depths are not as definite as well data. In a well, a formation may occur at exactly −4623 feet. But if a seismic time has been converted to a depth of −4623 feet, that is an approximation. In absolute depth, it can be 100 feet off, or more. Even in depth relative to nearby shot points, the −4623 feet can mean anything from say, −4610 to −4635 feet. One millisecond of error can mean from 5 to 10 feet, and it is impossible to pick precisely to the millisecond. So there is an amount of uncertainty that has been called seismic wobble. So, although well data can be, and are, contoured as precise numbers, with a contour usually placed exactly at its proportionate spacing between wells, the wobble of seismic data would make such precision not only meaningless, but misleading (Fig. 8-11).

So some kind of smoothing is required to achieve reasonable maps. There are various ways of bringing about this smoothing. Some people contour without paying close attention to exact values. That is, smoothness of contouring is allowed to take precedence over the last digit or so of the data (Fig. 8-12). A map so made appears to be incorrectly contoured. It isn't, but it is difficult to check.

Another way of contouring is to always put a contour between higher and lower data, but to pay no, or almost no, attention to its distance from any point (Fig. 8-13). This method is somewhat arbitrary, and the numbers which happen to be the same as the contour values are honored more than other numbers. The 500 contour has to go through a 500' datum. It has the advantage of being easy to check for errors, though. Just look along between two contours and make sure the data are all in the proper range.

My own way of contouring is a modification of this latter method. I assume that the value of a contour is slightly less than its label. For instance, I consider that a 500 foot contour is really about a 499.5 foot

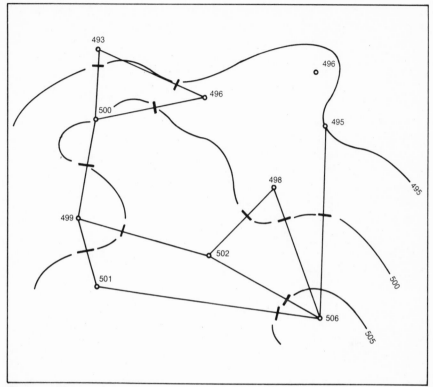

Fig. 8-11 Precise contouring

contour. Then I don't favor any data, as the label on the contour is not the same as any seismic datum. So the 500 contour doesn't have to go through a 500-ft datum. This is even easier to check, as all data, say from 510 through 519, are between the 510 (labelled) and 520 (labelled) contours (Fig. 8-14).

Another special characteristic comes from the way seismic data are arranged. Wells are usually scattered around the countryside, except when bunched up on a producing field (Fig. 8-15). Seismic data, though, are close together in long lines, but with relatively wide spaces between the lines, (Fig. 8-16). If the data were as far apart on the lines as the distance between lines, then there would be a good grid-like pattern for contouring. But to achieve this pattern, you would have to either shoot many more lines (Fig. 8-17), or throw away most of the expensive data (Fig. 8-18). Neither of these is economically sound. In a few areas

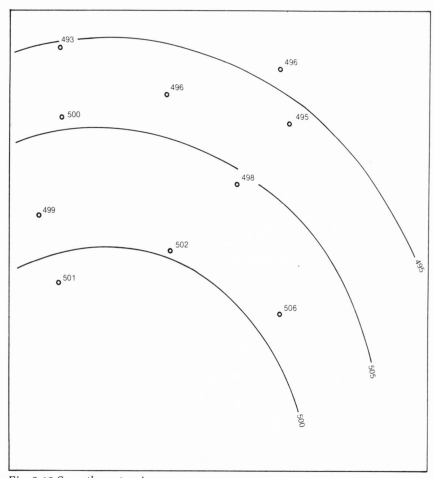

Fig. 8-12 Smooth contouring

requiring close control, the lines may be near enough to each other to produce a close grid of data. Throwing away most of the data could of course be used as a first approximation to contouring, to see what the trends are, before being influenced by the shooting pattern.

In normal contouring, this pattern of blank areas surrounded by dense control causes a peculiar phenomenon. The contours tend to follow the lines of control (Fig. 8-19). How did the subsurface know where we were going to shoot? In an area in North Dakota, I started a large project, in the winter. It was to be a reconnaissance program, with points all over. But

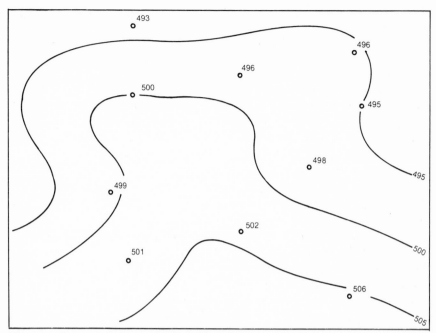

Fig. 8-13 Contours between high and low data

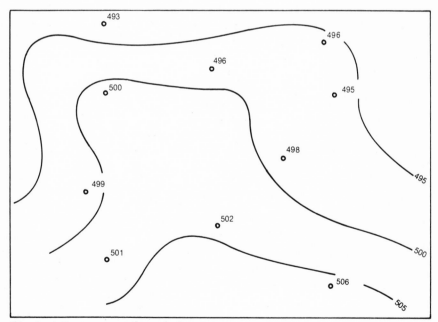

Fig. 8-14 Contours not through data

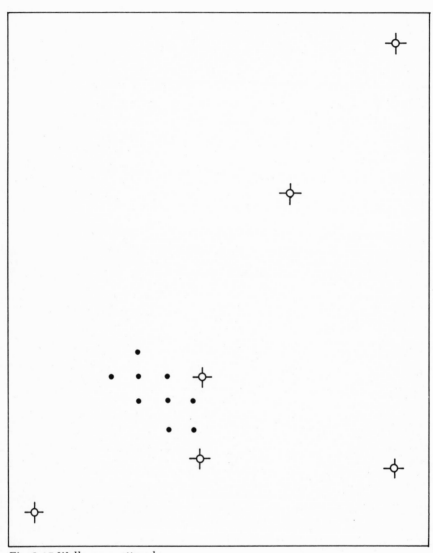

Fig. 8-15 Wells are scattered

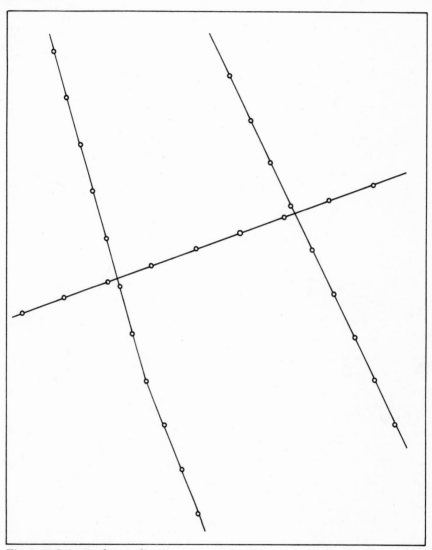

Fig. 8-16 Seismic data in lines

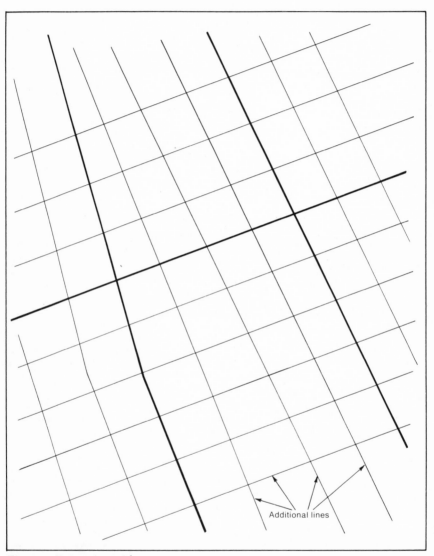

Fig. 8-17 Expensive grid

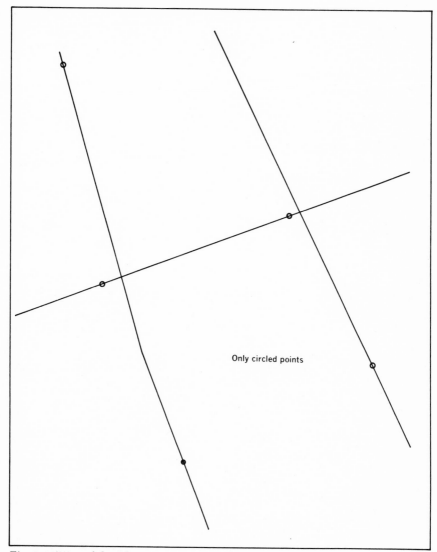

Only circled points

Fig. 8-18 Wasteful grid

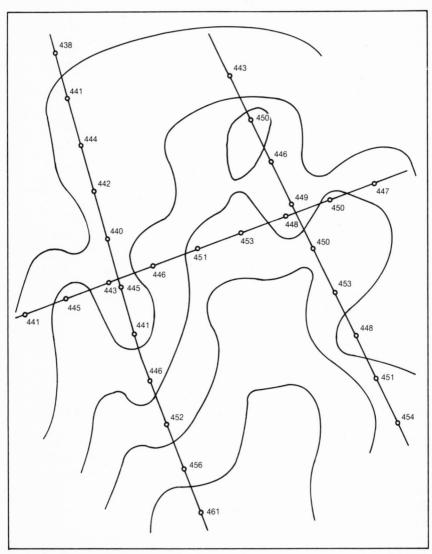

Fig. 8-19 Contours follow control

because it was winter, we were allowed to shoot the township lines first, and would go inside in summer, when it would be easier to get around. So we shot around two adjacent townships. What happened? The contours followed the township lines. So I had one 36 square mile high and one 36 square mile low (Fig. 8-20). If my contouring decisions had been different, the high and low would have switched townships.

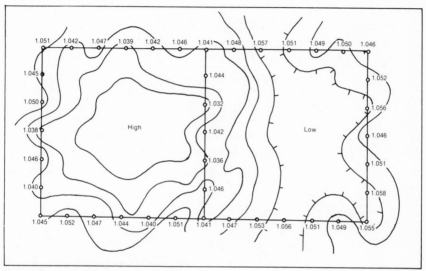

Fig. 8-20 High and low townships

You can guess what happened. Come summer, we shot inside. And, like most of the Williston Basin, bumpty-bump, the map looked like a tub of guts. No big features at all, (Fig. 8-21).

So what is to be done with a line or two of control, in an area of small features and low relief? Cautious contouring on the lines makes the contours swing back and forth across them. And seismic wobble adds to the effect (Fig. 8-22). Bold contouring, projecting trends from one line to another, becomes silly (Fig. 8-23). Drawing just short bits of contours across the line looks like railroad ties, and is hard to understand (Fig. 8-24). Somewhat better is to draw little round beads on the line, where the high data are, and hope nobody thinks they really are structures (Fig. 8-25). Probably best is a compromise between the last two, that is, the railroad ties, but curved somewhat, so the contours on highs bulge outward like parentheses (Fig. 8-26).

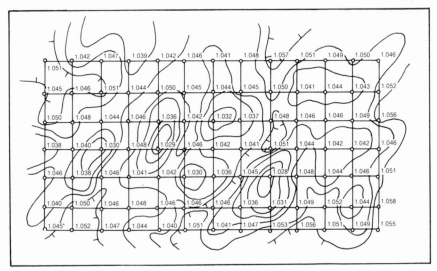

Fig. 8-21 No big features

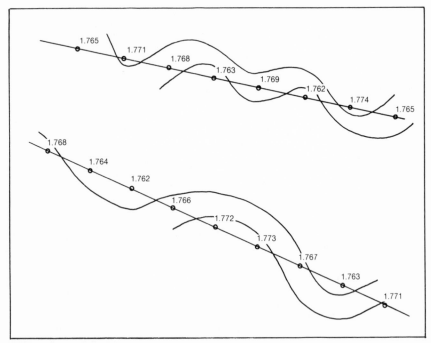

Fig. 8-22 Along control

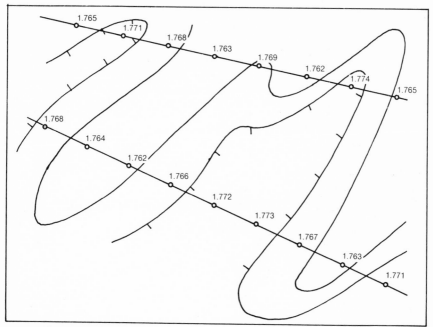

Fig. 8-23 Bold contours

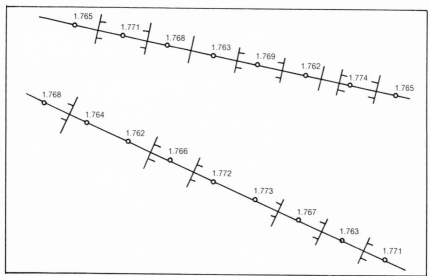

Fig. 8-24 Like railroad ties

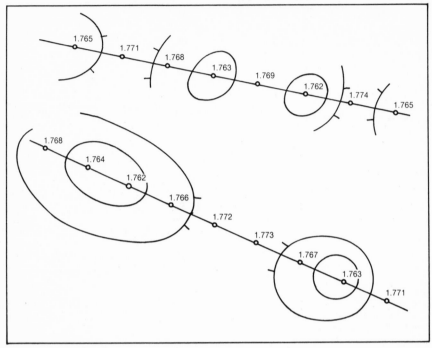

Fig. 8-25 Like beads

Computer Contouring

Data can also be contoured by computer. As with other computer operations, this is economically worthwhile only if there is a large quantity of data to be handled. Considerable time is required to put the data on punched cards or tape, and to get set up for the rapid computer operation to get started. Most maps of seismic areas require less time to contour by hand. Computer contouring is generally more applicable to gravity or magnetic data, which may be in much greater quantities than seismic data. Its best seismic application is in a large area with dense control, and a number of horizons, intervals, maybe alternate data made using varied corrections. The program could not only contour the quantity of data rapidly, it could also add data from new shooting and re-contour.

Another characteristic of computer contouring is uniformity. The same program will contour two sets of data in comparable ways, or do one set of data twice with the same result. This "impartiality" though,

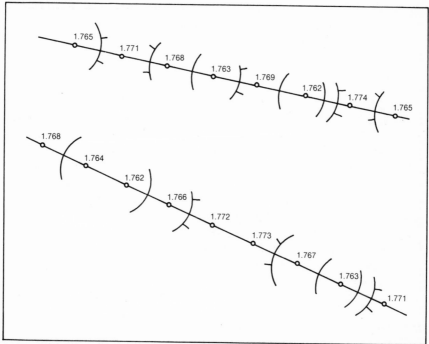

Fig. 8-26 *Compromise*

depends on the criteria put in the program. And writing a program is an attempt to put into instructions for a computer, the criteria a person uses when contouring. A different program will use a different set of "prejudices" in *its* "impartial contouring.

Input for computer contouring is: locations of shot points, in some convenient system, say distances north and east from a corner of the map; and data at the shot points for the horizon to be contoured.

It is easier for a program to handle data in a uniform grid than in the happenso arrangement of seismic lines. So in a typical program, the computer interpolates data at the points of a uniform grid. For each grid point, shot point data within a certain distance of it are averaged, with maybe those a little farther away included but given less weight. The grid point data are contoured. The contours are printed on paper or film, but not the grid data. Shot points and their data are usually included.

Computer contouring is a good way to get a first contouring of a mass of data. It can indicate the areas of regional dip, that do not call for further investigation. In the anomalous-looking spots, a geophysicist may re-contour by hand, making different choices at places that could be contoured in alternate ways, perhaps using concepts gained from other experience in the general area.

Time Interval Maps

Time interval mapping is useful in several situations. It has advantages similar to those of flattening record sections, naturally, as it is a related process. To make a time interval map, times for two horizons are picked at each shot point, and the smaller time subtracted from the greater. These differences, or time intervals, are plotted on a map and contoured. The map shows, in time, something like the thickness between the two horizons. It is a vertical measurement, except for the fact that the energy is reflected perpendicularly to the beds, which may not be horizontal. If there is steep dip, the map may not have much meaning. With a fairly flat area, so that problem isn't severe, the map has the advantage of not being troubled by near-surface irregularities. Any near-surface variation has the same effect on one horizon as it does on another, so does not affect the difference between them.

Time interval maps are particularly useful in detecting small or subtle features in areas that have glacial drift, lignite beds, or other difficult velocity situations near the surface. If the shallower horizon chosen is one that has very little dip, then not much error is made in assuming that it is flat. In this sense, the interval map may fairly well represent the structure of the deeper horizon.

There is a terminology confusion about this type of map between Americans and Europeans. Among Americans, the reasoning is that the map is like an isopach map, the "pach" part of isopach representing thickness. So, as this type of map is about the same, but in time, "chron" was substituted to make "isochron". In Europe, however, more correctly, "iso" for "same", and "chron" for "time" are put together to mean any map that is contoured in seismic time. A contour is a line along the same time. I confess to having been one of those who promoted the name "isochron" for an interval map specifically. So, partly to make amends, but, mostly to avoid confusion, I now speak very strictly of time interval maps and don't use the word "isochron" in either sense.

Accuracy vs. Correctness

How accurate is the seismic method?
People have told me they wished they could add some indication of

accuracy to their seismic data, like, I suppose, -3273 ft $\pm$ 50 ft. That would be nice if the seismograph was that good, but in practical use it just isn't.

The seismic wiggles lend themselves to picking only at the peaks and troughs. And adjacent peaks may be about 150 ft apart. So, if a formation you want to interpret is 20 ft thick, it is clear that these wiggles are a very crude measure. Granted, if the formation is a good reflector, then some reflection should about parallel it, but it still looks pretty crude.

Then, consider two wavelets starting at the top of that formation. Right at the start, there isn't anything that can be picked, so, to map it, something else, usually the first strong trough, is picked. To make this trough tie wells, some kind of fudge factor is needed. Often it is put in unintentionally. If you use the time to the reflection (the strong trough), and the depth to the formation in a well to get a velocity, then that velocity will not be quite correct, but will make the data tie the well. Or, if a velocity is known from velocity surveys in wells, or even from seismic data, then that velocity can be used. In all likelihood, the pick won't tie exactly, using that velocity, so, some number of feet or milliseconds is added to or subtracted from all the picks of that reflection, to tie the wells. Obviously, it must be consistent over the area, either constant or following some pattern, not just finagled at each well.

With all this jury-rigging, if there is a good, consistent reflection following a consistent horizon, the accuracy may be something like $\pm$ 50 to 100 ft, in terms of absolute well ties in depth.

In another sense, simple *relative* depth to determine whether some point is higher than others around it, this accuracy and a little better is fairly easy to obtain. That's nice, as it finds oil in some subtle situations.

But the big problem is accuracy vs. correctness. There isn't much gain in working on the minute hand of a clock, if the hour hand is broken.

The correctness problems of seismic interpretation generally involve staying on the same horizon. In crossing a fault or a patch of poor reflections, the interpretation may be incorrect by how much? 500'? 1000'?

Or, in a new area with no well control, identification of horizons may be off by almost any amount. In this case, velocities also may be pretty poor, particularly in the deeper part of the section, where seismically-determined velocity breaks down.

Sometimes, a well is drilled, and, by sheer happenso, the seismic prediction turns out to be off by something like 7 feet! The danger here is that someone in management will think this is genuine precision, and expect it on later wells. When congratulated for some accident like that, I have tended to say, "No, it was three hundred feet", and then go into a lecture on the crudeness of the instrument.

All this doesn't detract from the ability of seismic exploration to find oil. It finds most of the oil located at present. But unrealistic precision should not be expected of it.

±500 or 500±

There is a growing confusion about the use of the symbol "±". Perhaps it will be useful to consider the various meanings achieved by putting pluses and minuses before and after numbers:

+500
−500
500+
500−
500±
500±25
10000±500
+500±25
−500±25
±500

In other words, 500± means "about 500", ±500 doesn't mean that. It means either +500 or −500.

Bright Spot

The newest important advance in geophysics is the bright spot, hot spot, amplitude anomaly, or whatever other name may have been thought up for it.

As seen on a record section, a bright spot is just a reflection that is much stronger than usual for a limited distance (Fig. 8-27). Its importance is that it *may* indicate gases directly. A seismic reflection occurs where there is a velocity interface, that is, the sound is traveling through rock at the velocity for that kind of rock, and then encounters rock that transmits sound at a different velocity. This difference causes the echo, or reflection, and the greater the difference, the stronger the reflection. Also polarity is affected. A change from slower to faster will cause the trace to "break", swing one way, while a change to slower makes it break the other way.

So a reflection from the interface between two rock layers should look about the same from place to place, as long as the composition and

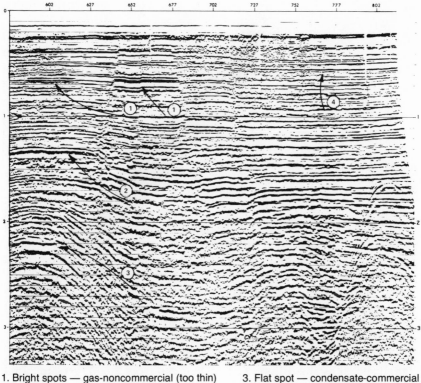

1. Bright spots — gas-noncommercial (too thin)   3. Flat spot — condensate-commercial
2. Bright spot — gas-commercial   4. Bright spot — Lithologic change,
                                                                                          not gas

Fig. 8-27 Bright spots and flat spot in offshore Louisiana

Credit: Seiscom-Delta Inc.

thickness of the layers remain the same, so other reflections won't interfere in different ways. However, porous rocks have fluids in them—water, oil, gas, even air, if nothing else. And these fluids transmit sound at different speeds. So the sound velocity of rock is affected by the fluid in it.

Now consider a porous rock layer with the pore spaces filled with the usual subsurface fluid, salt water (Fig. 8-28). The water has a velocity of about 5,000 ft/sec, while the rock, not counting the pore spaces, has a much higher velocity, say 20,000 ft/sec. The combination of rock and water will have a velocity somewhere between the two, depending on the proportions of rock and water, maybe 10,000 ft/sec.

Velocities generally increase with depth, so let's assume that the layer

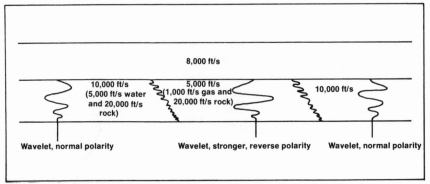

Fig. 8-28 Bright spot caused by presence of gas in rock

above it has a velocity of 8,000 ft/sec. Then the strength of the reflection is a result of that 2,000 ft/sec difference between the 8,000 and 10,000.

Also, the change is from a lower to a higher velocity, so the polarity is called positive, with the trace breaking, say, to the right. Now if, for some reason, some extent of the porous rock contains not water, but gas, the velocity relationship will not be the same. The gas may have a velocity of something like the sound velocity of air, around 1,000 ft/sec or, as it is compressed underground, somewhat higher.

The velocity and the rock's own 20,000 may combine to make, say, 5,000 ft/sec. Now the difference has changed from 2,000 ft/sec to 3,000 ft/sec, so the reflection should be stronger. Also, the change is now from a higher to a lower velocity, so the polarity is changed from positive to negative.

So the record section has a reflection that, for a limited distance, is stronger and has opposite polarity, so what is normally a peak becomes a trough. Then, it changes back to normal again. That is the most obvious kind of bright spot to look for—a briefly outstanding reflection. The polarity reversal may not be so easy to detect, as it may seem that the reflection moves a little up or down, instead.

Now if the velocities are different, all sorts of situations can occur. The velocity contrast can be less than normal instead of greater, so the exceptional part of the reflection is weaker. This makes, not a bright spot, but a dim spot. The velocities can change but remain in the same direction as normal, so the polarity is the same, rather than opposite. A trouble with a dim spot is that, although it may be just as significant as a bright spot, it may not show up as well. Bright and dim spots are spoken of together as amplitude anomalies.

A dim spot is particularly likely to occur in limestone, as the lime has a fairly high velocity. A limestone layer, or a reef, may thus contrast greatly with the overlying material, maybe a shale, producing a strong reflection at the interface between the two. If the limestone contains gas, reducing its velocity of sound, that may make the velocities of the lime and shale more nearly the same. So the reflection may be poorer, a dim spot. It may have the same polarity as the normal reflection, or, if the gas pushed the velocity somewhat below that of the shale, the polarity will be reversed.

Bright spots, in a different way, can show up about as clearly on regular sections as on true amplitude sections. If there is a strong bright spot on the data, then, in the processing to equalize the reflections, that one is toned down to more nearly the amplitude of the others, and of the same reflection in other places. But one reflection can't be reduced alone. After all, if the equalizing was perfect, there wouldn't be any reflections left on the section at all, just dead traces. So in reducing the amplitude of the strong reflection, the events above and below it are reduced to below normal amplitude. The attempt to equalize leaves the bright spot somewhat stronger than normal, and the others weaker. So there is a sort of shadow zone of dimness around the bright spot. This shadow is quite easy to recognize, and is as good a way to see a bright spot as on a true amplitude section. A true amplitude section, though, is more numerically correct, showing better just how bright the spot is.

Not much gas is necessary to change the velocity of a rock, so a bright spot may indicate very little gas, maybe as little as 2% of the volume of the rock. The bright spot tells us there may be gas, in a more direct way than seismograph does otherwise, but it doesn't explain whether it is in commercial quantities or not. And lithologic variations, coal beds, etc., may produce bright or dim spots in the same way gas does.

This discussion has been primarily about gas, as it has a much slower velocity than water. Oil has only a somewhat slower velocity than water, so it doesn't make as big a change in the velocity of the rock.

Careful analyses of amplitude anomalies can be made by specialists, calculating the coefficients of reflection at the various reflecting surfaces, to put relative amplitudes on a more quantitative basis.

Flat Spot

A phenomenon similar to the bright spot, is the flat spot, a bit of reflection that is flat, or unusually flat (refer back to Fig. 8-27).

What makes for a flat reflection? A flat velocity interface. Then what

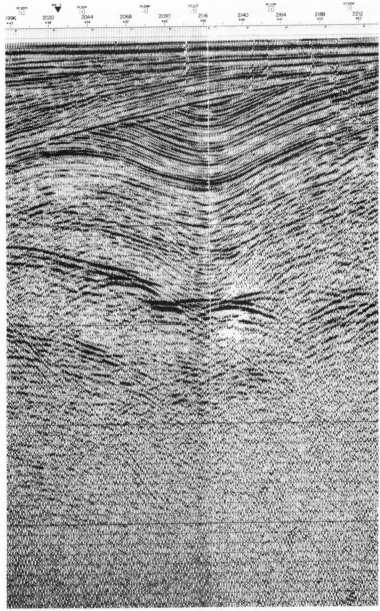

Fig. 8-29 Offshore Oregon-true amplitude section
                                            Credit: Western Geophysical Co.

makes a flat velocity interface? A flat-lying rock layer would, of course, but it would probably not be unusual for the vicinity, and so wouldn't stand out as having special significance. But there is one thing that is almost always pretty flat, the top surface of a liquid. So if there is a gas-water or a gas-oil contact, the top of the water or oil should be flat. If all the reflections around have some regional dip, and one flat reflection is among them, then it may well be a reflection from the velocity difference between gas and oil or gas and water.

An oil-water contact will also be flat, and may appear on the section, but the velocity contrast is much less, so the reflection is not likely to be apparent.

The ideal situation is to see a structure on the section, with a flat reflection below it. The structure may be filled with gas down to the flat reflection.

There may be more indirect relationships between a liquid surface and a flat reflection, as when cementation of porous rock has taken place, a mineral's coming out of solution in the water, but above, in the oil or gas, the mineral could not be in solution. So there can be a lithologic change at the liquid surface—present or old surface.

A number or flat spots on one or both sides of a fault, may indicate a seepage of gas up the fault plane, displacing water in beds along its way.

Some factors can make the flat spot not exactly flat. A lateral change of velocity in the rocks above, may give an apparent tilt to the reflection. Or, there may be a tilted contact, caused by hydrodynamic forces in the slow migration of fluids in the porous rock.

So the thing to look for is a reflection that is flatter than the other reflections around it, preferably with an abrupt start and stop at the ends, a reflection that does not look like it could be showing the attitude of rock layering on that section. Then it can represent the upper surface of a liquid, and the lower surface of a body of gas. The gas does not need to be composed of hydrocarbons, it can be carbon dioxide or any other exotic gases that are sometimes found in the earth.

Stratigraphy From Seismic Sections

Record sections are primarily good for showing structure, but, as they improve in quality through the years, the detail on them can show some stratigraphic features.

Deposition is shown in the way reflections conform to each other. A zone of parallel reflections indicates uniform deposition. Thickening in one direction can indicate deposition in a subsiding basin. An absence of reflections below a scarp may result from a quantity of sediment dropped over the edge, with little bedding.

Competence of formations may be indicated by minor faulting. If one horizon is broken by many small faults, while an adjoining horizon appears only to bend at those faults, then the first may be brittle and the second more plastic.

Misleading Features

Seismic sections improve in appearance with progress in shooting and processing. As they improve, they look more and more like a cutaway of the earth, with reflections having the configuration of the layers of rock. So, to an increasing degree, they can be interpreted by a person who is not trained in geophysics. This is one of the goals of the developments—to produce sections that can be readily understood, without the need for a specialist to interpret them.

But—that goal hasn't been reached yet. There are a number of features that appear on a record section, that must be interpreted with a specialized understanding of the way they are produced, features that can mislead, and have misled, people into drilling dry holes unnecessarily.

These features have been covered as the topics were reached in the book but will now be put together in one place.

Multiple reflections look like the primary reflections, but apparently deeper on the section. They dip more steeply than the primaries, so look like quite reasonable geologic features. They can be so strong they override the deeper primary reflections, so the deep structure is not seen. They can also make it seem that there is more sedimentary section than there actually is, by occurring on the part of the seismic section that represents primary positions below the basement. Stacking makes multiples less evident. So does deconvolution. But often they show through anyway, especially in the deeper part of the section.

Diffractions can look like symmetrical anticlines. Or, one side of a diffraction can appear to be a down-curving extension of a reflection. So, if a reflection is dipping down from a fault, a diffraction at the fault position can look like the reflection begins to dip in the other direction. Diffractions can be pretty well cleaned up by migration, so the more advanced type of section, the migrated one, is not so misleading in this respect.

Similarly, an unmigrated section may exhibit a buried focus, which may appear to include an anticline where there is actually a syncline. And ordinary dips, unmigrated, give a false idea of the amount of closure on a structure.

A dip that comes from a horizon out of the plane of the section, off to one side of the section, can, even after the section is migrated, be confused with reflections from below the line of shooting. Three-dimensional migration is the most complete solution of this problem. If it is not available, or not possible with the data at hand, judgement must be applied to determine whether there are features to the side that might be recorded on the section. If dips conflict with each other, dipping in opposite directions and crossing, it may be that some are from one side.

A large feature on the surface, river bank, cliff, or the like, may reflect some seismic energy for a long distance. The appearance of this phenomenon on the section is that of a long straight dipping reflection, that, extended to the surface, meets the surface feature. It may cut across other reflections. Once recognized, it can be ignored.

Velocity anomalies appear to be structural, but are caused by the sound going through salt, reefs, igneous features, and distorted bedding. Identifying the source of the anomaly can allow the distortion to be recognized. In some cases, velocity data is complete enough to permit a corrected section to be made, so the velocity anomaly is removed.

Pinchouts are not where they appear to be on the section, but somewhere beyond the area where the two layers can be individually recognized.

Most seismic sections have considerable vertical exaggeration, as well as a non-uniform vertical scale—in terms of depth. The scale is uniform only in terms of time. The scale should be taken into account in thinking about the shape of a feature and the angle of a dip or a fault. The vertical exaggeration is generally necessary to make subtle features noticeable.

Both the vertical exaggeration and the non-uniform scale can be eliminated, if velocity information is adequate, by making a depth section at a one to one scale. Or a depth section can be made at any selected vertical exaggeration. But a caution is necessary. Velocity information usually is not adequate to make a trustworthy section in depth.

Interpretation Sequence

The actual interpretation process starts when sections are received by a geophysicist. His part in the operation, and the interpretation he makes, are improved if he is in at the start of the project, and works on the determination to investigate the area, program planning, field work, data processing. Decisions all along the line can have a major effect on whether oil is found as a result of the interpreting.

But he starts to actually interpret when he has sections to do so with. Areas and problems differ, but a more or less typical process of interpreting goes like this.

The interpreter looks over the sections for the better and more continuous reflections. Then he tries to identify them. If a line goes by a well with a synthetic seismogram, or if the area is well known to him, identification is probably easy. If not, he uses various tricks to find out what formation is represented by a certain reflection. Notice that you go from reflection to formation, not the other way around. It would be nice to decide you want to map a particular formation, say the pay zone in the area, but if there isn't a reliable reflection from it, you can't do it. You can select a good reflection that looks like it is control for that formation, i.e., like it will pretty well parallel the formation. Then the reflection can be mapped, and used to represent the formation of interest. A correction factor can be applied to make it fit wells as though it was the formation. Often, maybe usually, the problem doesn't exist, as there is a reflection from the formation of interest. But it's worth being aware that you can't always map whatever formation you want.

Having the reflections to be mapped selected, the interpreter starts picking. He may color a reflection with colored pencil. He times the section, i.e., marks the reflection times at points under the shot points. He may read the times direct from the timing lines, or may use a scale cut from the edge of a section, or drawn up for the purpose. If a scale is used, he puts it in register with a nearby timing line, so paper stretch of either section or scale will not throw the numbers off. The reflection times may be written on the section, or tabulated by shot point number, or put straight on a map.

At some point, the section may intersect with another line. The picking may be carried on past the intersection, or may turn and go along the other line. Eventually it must go both ways. Picking around loops, sooner or later some loop will fail to tie. The mistie must be somehow resolved. Then there will be other problems—a loop that ties, but must be undone so another can tie, faults with indeterminate throw, etc. Some interpreters just color the reflection, not timing any picks until the whole area has been made to tie. Others mark times on a map as they go, so they can see what is happening overall on the map, or so there can be interim mapping to show prospects, meet deadlines, etc.

The map, after all the picking is resolved, has times at the shot points. Usually not all shot points are timed, but every fifth, tenth, or something. There are often fault indications on the map. There may be only marks to show where a fault crosses a line, and direction of throw. Amount of throw, if it seems reliable, may be added, or reflection times against the fault, again if reliable. If it is possible to correlate a fault from one line to another, that correlation may be indicated.

The map is contoured, and the faults drawn in. Both contours and fault connections from line to line are ways of displaying best guesses where there isn't data. So they can be trusted more on a line, or where there is a concentration of lines, and less where the lines get sparser.

From the contoured map, prospects or locations for drilling are picked out. A closed high feature, or contours closed against a fault, may show a place to investigate further with more shooting, or to drill a well.

Non-structural features may be added to the map, or put on a separate map. Bright spots, flat spots, reefs may be prospective.

Interval maps may be made, by subtracting the times on one map from those on another. These can show reefs or pinchouts, or substantiate the validity of the horizon maps. If near surface corrections are difficult, the interval maps, being free of near surface effects, will not have the problems caused by them.

Features, especially those recommended for drilling, are discussed with geologists to consider the features in the light of both disciplines.

9

Refraction Exploration

Refraction

Refraction of sound plays a large part in seismic exploration. It dictates the paths of energy in reflection work. It helps make near-surface corrections for reflection records. And there is the refraction method of exploration, as distinguished from reflection.

The principle of refraction is that energy going from one medium to another that transmits it at a different speed, is bent. Light, sound, radio waves, all act this way. The best-known example of refraction is the stick poked into water, and looking bent at the surface of the water (Fig. 1-27). Light travels at different speeds in air and water. There is also *reflection* at the surface, so sometimes you can see in the water, both the refracted lower part of the stick and the reflection of the upper part. The reflection you see is the light that went from air to water and bounced back, while the refraction is the light that went from the water into the air and was bent.

Seismic energy, traveling down into the earth, is similarly bent at velocity interfaces that it does not strike perpendicularly. There is a definite, calculable, amount of bend for each velocity contrast and angle of approach to the interface. When there are a great many interfaces, then the energy path, making all those little bends, becomes nearly a smooth curve. When the velocity increases steadily, as with compaction of a shale with depth, the bending does become a smooth curve.

The other uses of refraction—for near-surface corrections and for refraction surveys—depend on having the receivers considerably farther from the source than the depth of the horizon.

From a seismic source, sound goes out in all directions. Every bit that strikes an interface, sets it to vibrating, so it is as though there was a new source there. So sound goes out in all directions from each of those points. The part that goes upward is seismically useful, as it can be received on the surface. In general the sound that arrives at a point by way of a horizon is the part that takes the quickest path from source to horizon to receiver. If the instrument is near the source, then the sound to reach it by the quickest route from a horizon is that which was reflected back from a point midway between source and receiver. However, if the receiver is far from the source, and if, as is usually the case, the lower layer of the interface has the higher velocity, a quicker path will be down to the faster layer, along it, and back to the surface. This is like the decision when walking across a plowed field. If there is a road nearby, parallel to your course, then, depending on the distance you are going and how far out of the way the road is, at some point it is quicker to detour over to the road, than it is to walk straight across the field.

Look at what happens when the refraction path is quickest. There are groups of geophones, spread out in a line. Each group receives energy that has gone down to a fast layer, along it, and back up (Fig. 9-1). For all the groups, the route down is the same one. And the paths up to the surface are about equal to each other. Topography may make them different, but that can be corrected for. So the up and down paths are about the same, but there is a difference from group to group in the time the energy spent traveling horizontally in the faster layer. What this all amounts to is that the time it takes for the sound to go the known distance from one group to the next, through the high velocity layer, gives the velocity of sound in that layer. Divide distance between groups by time between traces to get velocity. If the times are plotted vertically on a

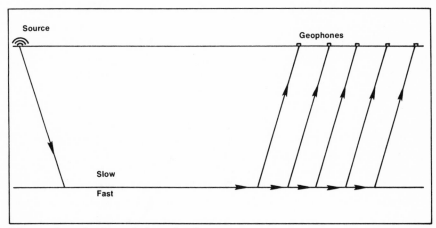

Fig. 9-1 Sound refracted along fast layer

graph against horizontal distance, a slanting line results (Fig. 9-2). As time in the fast layer is the thing that was measured, then the slope of the line represents how fast the sound went through this layer. The intercept—the height of the lowest point of the line, where it is at zero distance—represents the time the energy took going down and up. So the slope is the velocity in the fast layer, and the intercept, the time spent in the slower zone.

This applies only if the horizon is flat. If it has dip, then the dip will also affect the slope of the plotted line. To get around this problem, refraction lines are also shot in reverse, and the slopes of the two lines averaged.

This kind of plot is the basis of both uses of refraction. For near-surface corrections, the regular geophone spread for reflection shooting is spaced for obtaining reflections from the depths of interest. But the base of the weathering is so shallow, that for it, most of that same spread is properly spaced for receiving refracted energy. So this near-surface refraction gives the velocity of the sub-weathering layer, and aids in determining depth to that layer (Fig. 9-3). The refracted energy in this case is the first energy to arrive at the farther traces, so is called first breaks, or first kicks.

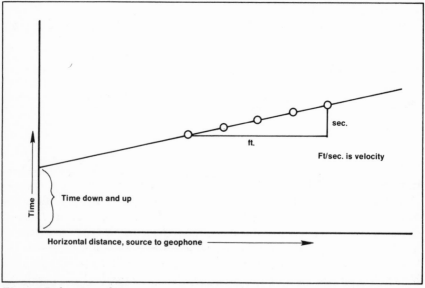

Fig. 9-2 Refraction plot

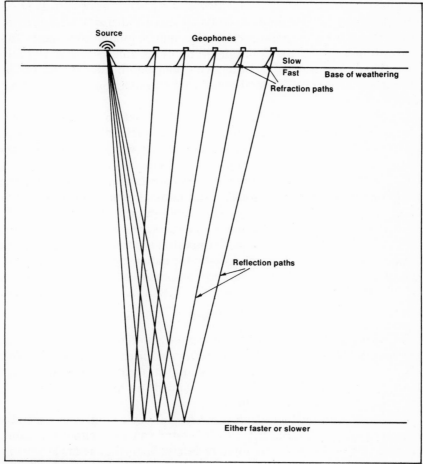

Fig. 9-3 Refraction in reflection shooting

The very same geometry, but on a larger scale, is used in refraction exploration. In order to have the instruments far away in proportion to the depths of mapping horizons, they are placed several miles from the shot point. A refraction crew, because of this greater distance, has its own special problems of logistics. More vehicles may be needed than on a reflection crew. Communication between units is by radio.

From one to several horizons will be mapped from the refraction data. The velocities of the formations themselves—not average velocities of everything above them, as in RMS velocities from normal moveout—are obtained. This is one of the advantages of refraction shooting. There is little trouble with continuous identification of a horizon. After a gap in data or a fault, you can get back on the right refraction reliably. Its big disadvantage is that the depth determinations are much less accurate than those in reflection shooting. The poor depth information makes people use refraction, generally, only when reflection won't work for some reason.

Refraction was the first type of seismic exploration to be in common use. Salt domes were found by refraction before reflection became popular. The higher velocity of sound in the salt was detected on lines radiating from a point in a fan shape, therefore called fan shooting. A quicker energy reception on one line than on another indicated that the first might have traveled through a salt dome. Refraction is now sometimes used in mountainous regions where terrain is too rough to allow occupying all the positions wanted, so it can survey under the obstacles. It is often used in virgin areas, with the velocity information serving as a rough identification, if not of specific formations, at least of the larger geologic units—Cretaceous, Tertiary, basement, etc.

Sonobuoy

There is a simple means of obtaining refraction data economically in the course of an offshore reflection survey. In the course of the reflection shooting, the boat usually runs along straight lines, trailing its geophones (hydrophones) on a cable, and firing its shots (airgun, gas gun, etc.) at frequent intervals.

The description of refraction shooting utilized a single source and a spread of geophones. But the paths of sound can go in either direction. Source and receiver are completely interchangeable, as far as energy path diagrams are concerned. So one geophone group and a number of shots would give the same information as one shot and a number of geophone groups.

Taking advantage of this interchangeability, and of the shots that are being fired anyway, a sonobuoy is used. It is a device consisting of a geophone or geophones built into a plastic tube about three feet long, that floats upright and carries a radio transmitter. The sonobuoy is thrown overboard. The geophone is released to hang suspended at the end of a wire, maybe 100 feet down in the water. Sea water activates a battery, and the radio starts broadcasting the output of the geophones. The broadcast is received and recorded on magnetic tape on the boat. As the shots continue and the boat gets far enough away, the geophone receives refraction data. The sonobuoy is designed to sink in a few hours.

So the refraction information is a sort of byproduct of the reflection survey. It also has some of the disadvantages of a byproduct. The sonobuoy is not in a fixed position, but may drift with wind or current. The line is shot in only one direction, not reversed. The shots are normal strength for reflection, and may not provide as much energy as would be best for refraction. But within these limitations, quite useful information on velocities can be obtained, and tied in with the reflection data.

Old Data

Availability of Old Data

Shooting is normally done by a contract crew working for a client, and the data belongs exclusively to that client; or by a company crew, so it belongs to the company. Seismic data is usually highly confidential. This secrecy has been broken down a bit by the idea of trading data. Companies are willing to let out data from areas in which they no longer have an interest, in exchange for some in an area they'd like to know about. The trading is cumbersome, and takes up a lot of employee time that could better be spent hunting oil instead of hunting old records and getting agreements signed, etc.

So the trade service came along. It is an organization that handles the leg work of trades for companies—and sometimes even stores the records so the company doesn't need to dig them out. The trade service became a library. It distributes maps of data available from different companies, combined onto one map, and arranges the trades so that very little company time is used. Joining the library may require depositing some set amount of data, with this deposit allowing you to take out from 2 to 4 times that amount.

The library is fine among large companies, but smaller companies without a backlog of data are pretty well left out. And often they are the very companies most likely to find the old data most useful.

So the library shifted emphasis from trade to sale of data, and the advantages of money over barter were discovered anew. A company can offer any amount of old data that it would like to earn some money from, and any company can purchase data without having to have records to deposit.

All these arrangements exist now, but the sale of data has become the most common. The expressions "trade service", "library," and "data exchange'' are used to describe the companies handling both sales and trades of data.

Working up prospects in areas that have had shooting done in the past, can be made more economical by using some of the old data.

Much of the old shooting is available for trade or purchase. It can be obtained by dealing directly with the company owning the data, but is most effectively handled through the data exchanges. Their maps of data available from a number of companies, make it easy to determine which control covers the area of interest, simplify the putting together of a combination of parts of different companies' shooting to make a survey covering your area.

Trade or purchase of old data usually gives you a copy of the records or sections, and the right to copy the tapes, if any. But it is stipulated that the data is for your own use only, and not for any other company. However, this is customarily interpreted to mean that data can be shown to partners in drilling deals, etc.

Price of old data varies with the type of data, and the company offering it. A very rough rule of thumb is that the price is on the order of a fifth of the cost of the shooting.

Value of Old Data

A basic question about old data is whether there is any point in bothering with it at all. Didn't the company that shot it squeeze all the anomalies out of it, and drill the good ones? For that matter, is there any reason to even look for oil in an area that somebody else has already shot? Are you that much smarter than all these other exploration people?

As for looking in an area that was already shot, new fields are constantly being found where people have been looking for a long time.

On using the old data, you don't have to be smarter to get something useful from the data. First, it gives you regional information. Then, your outlook isn't the same as the other company's. The leases you're concerned with aren't the same as theirs, you may be thinking of a different formation, or looking for smaller features or a different kind of feature. Putting together parts of two or more old surveys will give coverage that none of the companies had after shooting their own prospects. And last, maybe you *can* be smarter. Did anyone ever refuse to look at an electric log because other geologists already had that log available?

Old data can also be reprocessed to yield better information than before. The most striking example of this is from transcribing old paper records onto magnetic tape, so they can be processed digitally into record sections.

A remarkable change can take place in going from paper records to record sections, and a much better interpretation can often be made. The record section is almost required for display to management or to other companies.

Some other reasons for using old data are:

It is non-seasonal. You don't have to wait for a freeze-up, or dry season, to acquire it.

It is really turnkey. You know exactly what it will cost.

Sometimes it is the only kind of seismic data that can be obtained in an area. If shooting is no longer possible for some reason, the old records can still be used.

In most cases, the old shooting isn't expected to determine a well location. The lines probably weren't shot just where you need them, and any hitherto-unnoticed lead must be followed. So there will normally, unless the old work condemns the prospect, be some detail shooting to be done, or even a repeat of a key line or so, to verify the prospect with modern techniques.

But the old pattern is often about as good for reconnaissance as the one you'd lay out at the start, and your detail will go where you want it anyway.

Forms of Old Records

The advances in geophysics cause not only the quality, but also the form of data to change. So, if old data is to be worked, either from your company's files, or obtained by trade or purchase, it may be in any of several now obsolete forms.

For most of the development of seismic exploration, the data was recorded on individual paper records, each shot on a separate piece of paper. The very first had two traces on a record. The number of traces increased through the years. At different stages, the most common form had six, then twelve, 24, and 48. The records were made photographically, in a little darkroom in the recording truck. Three stainless steel pots were used, for developer, fixer, rinse water. To speed things up, the developer was used so hot you could just barely put your hand in it! Also, by the end of the day, the rinse water, which is intended to remove fixer, had become a fairly good fixer solution itself. With this kind of processing, it is no wonder that black spots from overdevelopment, black splotches from someone's inadvertently opening the door, brown stains from fixer remaining on the record, are all to be found on these paper records.

For a while they were dark with white lines for the traces. Photographic paper is a negative material, becoming black where light strikes it. These dark records were made by shining a light on the paper as it moved past the instrument. Wires wiggled in response to electrical signal from the geophones. The shadows of these wires produced the white traces. A special point in developing a record was to make sure the dark background was not too dark. It had to be a gray that was dark enough to contrast well with the white traces, but light enough that pencil marks could be seen on it. These were mostly pre-1940 records of the 6 trace era. You aren't likely to see many of them.

Then the mirror galvanometer was developed. It too had the fine wires that wiggled in a magnetic field, but the wires had tiny mirrors glued onto them. The mirrors reflected light instead of casting shadow, onto the paper, which was otherwise not illuminated. So white records with black lines were produced. These are the common paper records that might be encountered. It was still photographic, so still subject to light leaks, overdeveloping, staining.

With all paper records, the record was the only information obtained from the shot. After the shot, there was a paper record, and that was all. No chance to improve it. So the observers had to do all the data processing by setting their instruments before the shot. An observer shooting an area, tried several filters to eliminate undesired frequencies, at the start of the area, then used the filter setting that looked best, for the rest of the area, or sometimes "played it by ear" all through the area. Also, "mixing" was used. This was a combining of two or more traces to produce one trace. So adjacent traces were made to be more like each other as a means of smoothing to improve signal to noise ratio, under the assumption that the reflections of horizons would generally be nearly flat. This mixing, though, was detrimental to reflections from dipping horizons, or, if the surface of the ground was hilly, to all reflections. Mixing gradually went out of style, with records often using only a "very light" mix. Some records were shot with a dual presentation on the same piece of paper, one set of all the traces, mixed, the other unmixed. The decline of mixing was so gradual for primarily commercial reasons. It was difficult for a seismic contractor to sell its services with unmixed records if its competitor's mixed records looked so much smoother, and therefore, apparently better.

Paper records were generally two to three feet long, and tended to become wider through the years, progressing from six inches to a more or less standard eight, with some ten inches wide.

On a paper record, normal moveout makes the reflections curve, as the geophones farther from the shot receive the reflections later than the closer geophones. So the "near traces", the ones on each side of the shot, were the ones used for calculations of time and depth. The traces affected more by normal moveout were used just to follow continuous reflections from one record to the next. And, if the two spreads away from the shot point were of equal length, the two far traces could be used to determine the dip across the record.

On paper records, near surface corrections were necessarily made only as calculations. The traces could not be moved. Most corrections were simple enough for the calculations to be performed right on the paper record. Then it was always available for checking, referring to, changing as needed. As the records were interpreted, horizons were picked in pencil, reflection times marked, and corrected for near surface variations. So, normally, there are two numbers, the "raw" time and the corrected time, at each reflection. Then, if someone else reinterpreted the records, his picks and calculations were added, sometimes in color, to distinguish from the earlier marks. A still later interpretation would call for another color, or a distinctive mark.

The records were interpreted by laying them side by side and marking a pick from one to another. Later, it would not be obvious which reflection was which, so distinctive marks had to be made to distinguish between reflections. In an area in Southern Louisiana, a certain reflection was marked on the records in an area with a little circle in blue pencil. The crew, therefore, called the horizon the "blue o". And as they were in a largely French-speaking locality, the spelling was made French-looking. The horizon became the "bleueau".

So paper records will usually be found to be very marked up with pencil. People felt that the marks were the only way to reconstruct an interpretation, so all marks already on the record were considered as data that should not be removed.

Many of the paper records have been microfilmed. This has been for several purposes. As long as there was only one copy of a record, the data was very vulnerable to loss from fire, shipping error, wear and tear, etc. A microfilm copy, stored in a different place, was insurance. Also, the deterioration from staining meant that the paper records would keep getting worse, as time went on. Even though microfilm is also a photographic product, it is processed under much better conditions, with long life particularly in mind. Then the trade services needed a way to keep records on file without depriving the owners of them. So most data exchanges have a file of microfilms of paper records. If you buy a set of records from them, it will probably be in the form of enlargements from microfilm, blown up to either original size, or to a scale you choose. And, during the stage of deciding whether you want those records, they can be inspected for quality by looking at the microfilm on a reader.

Some paper records have since been transcribed onto tape, by following the wiggles with a sensor that produces electrical impulses which are recorded on tape. Then the records can be in the form of tape and record sections (Fig. 10-1). The filtering and mixing cannot be removed, but other processing steps can be taken. Traces can be corrected by moving up or down, normal moveout can be removed, the records can be assembled on a record section. And the section will not have light leaks, overdeveloped areas, or fixer stains. If any of these blemishes completely obliterated the traces, then there will be blank spots on the section.

Analog tape was developed and rapidly replaced paper records as the recording medium. The visual form became record sections, which were in a form very like present day sections. They tended to be rather large scale, in imitation of the paper records. The earlier analog-produced sections pre-dated the common depth point technique, so they are single fold, 100%, sections. So they didn't have the multiple attenuation of CDP sections or the improved filters of digital sections. In general, they had more noise in proportion to signal. Their form was the same, but they weren't as good. In a good record area, this may not matter much, and they may serve as well.

The analog tapes themselves though, were in a quite different form from digital tape. Whereas many shots are recorded on one reel of digital tape, in the case of analog tape, each shot was recorded on a separate piece of tape. The tapes were from 4 to 10 in wide and 2 or 3 ft long— rather like the dimensions of paper records. They were filed flat in flat boxes, with each tape in an individual envelope, or were rolled up in little plastic jars. A tape was put on a drum in the recording truck, the shot was fired and recorded, and the tape removed, annotated with shot point number, date, etc., and refiled. The tapes were similar to records also in having all the traces side by side. Twenty-four traces called for twenty-four channels on the tape.

Some analog tapes have since been transcribed into digital form, in which case they will be on a regular digital reel. This gives some of the advantages of digital records, but not all. The data is still limited by its being recorded in the field in analog form. The frequency and amplitude data aren't as good, for instance. But, on digital tape, digital bandpass filtering applies and deconvolution, migration, etc. And the processors don't have to keep putting tapes on drums and taking them off.

When digital tape began to be used, the forms of both sections and tape became essentially what they are today. But there have been advances all

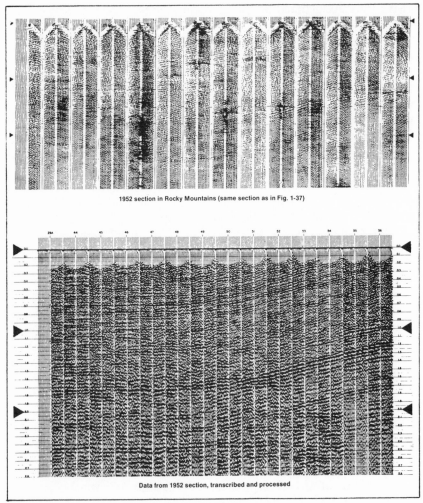

1952 section in Rocky Mountains (same section as in Fig. 1-37)

Data from 1952 section, transcribed and processed

Fig. 10-1 Transcription of old data

Credit: Geoscience Technology Services

along, so sections made 3 years ago may be considered old data and not
as good as the new. In some respects, even last year's shooting is pretty
dated. However, some of the advances have been made in field recording, and some in processing. The processing advances are largely available for the older data by reprocessing.

Offshore records shot with dynamite in the pre-tape, pre-stacking
period were, like the land records, shot as split spreads. But the records
from offshore have erratically lopsided looks to them, with the first

breaks indicating that the spreads are unequal, and the inequalities varying from record to record.

This is a result of the way they were shot, in the typical two-boat operation. The recording was done on a boat that also trailed the cable. The center of the cable was marked with a buoy, often two inner tubes. A separate shooting boat moved in to a position determined by eyeball to be offset from, but even with, the center of the cable. The charge was dropped overboard, suspended from a float, and connected by a firing line to the shooting boat. The boat moved away, and the shot was fired.

This system didn't get the shots just where they were wanted, so the lopsided-looking records were the result. The "off center" problem can be solved by plotting the two center traces and their first breaks on a graph, using a constant water velocity of 4900 or 5000 ft/sec, then making appropriate normal moveout corrections. If the records have been transcribed onto tape and made into a record section, then this has already been done by somebody.

Combining Old and New

Once, a new boss of mine, who had been out of seismic exploration for some years, asked me to tell him the new developments. There were very few. In fact, I was able to tell him of only one change that he hadn't heard of, and it was minor. That is hardly the case nowadays. Six months out of date is behind the times. The change came with the introduction of record sections and, shortly thereafter, magnetic tape recording. Since that time, advances have been both rapid and accelerating. There is a tendency to feel that only new data has value, and that old data is so much poorer that it should not be used. By "old data", people mean something like 3 or 4 years old.

The advances are rapid, and the philosophy of using only new records is, in part, valid. It is often so much better that old data just is not a substitute for it. In areas with severe multiple problems, or in amplitude (bright spot) investigations, the old data may not contribute much, and may have to be reshot, in critical locations.

However, as a matter of personal philosophy, I feel that data is too expensive and hard to come by, to be thrown out. In general, rather than reshoot an old line, I would shoot a new line far enough from the old one that both can contribute data to a map. A line through a well, or through a location of a planned well, should be the best available. If there is an old line through that location, then maybe a new line can go through the same spot, but in a different direction. One situation in which a new line *should* be a duplicate of an old one, is when it is intended as a check on

the old one, to see to what degree it can be believed. Then the duplica-
tion of one line may serve to calibrate a number of old lines, so the
information they do have can be incorporated with the new data.

The above sounds like new data is superior just because it is new. We
all know that, in general living, newer may not at all mean better. Well,
in seismic data too, new does not necessarily mean better. But, improvements in equipment and techniques are so rapid, that new *almost*
always does mean better, and a good deal better, at that.

Back to philosophy of use of old data. The natural, minimal effort, way
to work is to map the lines as they are shot, so the first shooting in an area
forms a framework for the interpretation. As new lines are added, they
are fitted into the framework, adding detail to the interpretation. This is
a good plan for work that is continuing, a few detail lines now, a few a
little later, as may be the case in some areas, particularly on land.
However, when shooting programs are large, and not so frequent, as in
many offshore areas, the developments in seismic exploration come into
the picture. Then, especially when the interval between programs has
been more than a year, it may be advisable to use a different approach,
with the new work interpreted, and used as a framework, and the old
adjusted to fit it. In other words, if some of the data has to be hammered
into place, let it be the poorer data. Why force better data to fit a
framework of poor data, just because the poor is already interpreted?
Well, there often *are* reasons, of course—deadlines, manpower shortag-
es, expense, other projects that need to be worked on.

And, if anybody wonders whether data really needs to be hammered
into place, well, take it from me, it does. This seismic data isn't that
absolute. It is often ambiguous, confusing, misleading. As in the saying,
we're a long way from our work. We're on top of the ground, and
working on a place a mile or so underground. We just sent a noise down
there and hoped it would bring reliable data back with it.

So, when there is a long time between the gathering of the old and new
data, and when there is time and manpower for it, and when the new
data is extensive enough, it should be interpreted first, and used as a
skeleton, with the old data made to fit it, or if parts won't fit, those parts
thrown out.

However, if some special aspect of the sections is being mapped, and
that aspect does not appear on the old data, then the old data can't be
incorporated in that map. For instance, if a bright spot map is being
made, and old data does not have true amplitude preserved in it, then it
cannot be added to the map. However, an accompanying structure map
may use both old and new data.

11

Types Of Areas

Large Structures

Large structures, either high-relief or broad and gentle, have their own subtleties that make them other than the cinch to find that they seem.

The broad low-relief structures can be spread over so much country that you shoot prospects on them, but don't find them because you never get off of them. This happened to a friend of mine in the Scurry Field of West Texas. Shot a whole prospect on top of it, but didn't have any lines extending beyond the feature. The field was discovered later, by someone else.

Similarly, a prospect may be on what seems to be regional dip, but is on the flank of such a feature.

The low relief on a broad structure may fail to excite just because not much action appears on the records.

The very steep structures are necessarily smaller in area. If they were both big and steep, they'd run out of geologic section, and have to stick up in the air. Some do, of course, for instance the sheepherder anticlines of Wyoming, so named, I would guess, because the sheepherders knew about them before the oil business got interested.

These big structures, particularly when they don't show on the surface, can be subtle because they are so steep. For energy to be reflected back to the shot point, and be perpendicular to the bed, the shot point must be pretty far from the structure. Then the wild-looking dip conflicting with more normal horizons from nearby, is easy to ignore as noise. Right over the structure, the turnover at the crest will show, but may extend only a short distance (Fig. 11-1). One prolific structure in Wyoming is producing because a seismic contractor saw one strange dip on a record, and insisted that it be investigated further.

Looking for these steep, narrow structures calls, first of all, for an

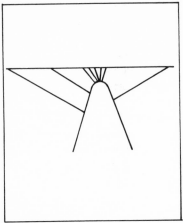

Fig. 11-1 Steep structure

open-minded attitude toward stray data, and second, for good migration of dips.

Small Features

Small features, of whatever geologic type, have some dominant exploration requirements, just because they are small.

Exploration has grown up based on a search for large anomalies. And in any new area it grows up again in the same way. Of course, bigger fields seem better, but the smaller, in area, ones are often well worth seeking, too. Some of the world's giant fields are very small in area.

Because of the large-anomaly trend, small fields tend to be looked for with big-field philosophy. This is like trying to catch minnows with a salmon net.

Some special rules for small fields are:

Seismic lines should form a fine enough mesh to find them. In some areas, this requires lines as close as a quarter mile apart. Widely-spaced lines are almost useless for this purpose, giving leads only to those features they happen to cross.

A well based on the seismic data should be drilled *only on a seismic line*. Off to the side of the line you can only *guess* that the location is on the feature.

From this it follows that the lines should run through locations that can be drilled. It is a waste of money to shoot lines on section line look in Montana, and Hummingbird in Saskatchewan. A technique appropriate to that type of area can probably find much more oil in these and other areas in the basin.

The salt solution idea is described thoroughly in the Saskatchewan Department of Mineral Resources publication, Report 76, which goes into it in detail for the Outlook area. Briefly, salt dissolves locally, allowing later deposition to fill in, becoming thick where the solution took place. Then, after more deposition, the rest of the salt layer is removed, so the overlying beds sag. The thick spot where the first deposition took place causes beds over it to drape, creating a trap. The features resulting are small, say from 640 to 160 acres in size. But there can be a number of them in a small area, so they are worth looking for.

Exploring for these features requires close seismic control, like any other small features.

Prospects are found largely by interval mapping. The beds under the feature are undisturbed, and over it there is drape, which diminishes toward the surface. So intervals from whatever is the pay zone up or down will show the feature by thinning or thickening.

The near-surface situation may be complicated by glacial drift. This is not a great problem though, as any feature found on a structure map can be checked on an interval map.

Salt Domes

Salt domes are the key to petroleum accumulation in many areas, including the Gulf Coast, North Germany, parts of the Canadian Arctic.

Salt dome exploration was the first common exploration use of seismograph. Refraction lines were shot so the energy, where it came in more quickly than normal, indicated a mass of salt. Reflection shooting is now used to find salt domes and determine their shapes.

A round topographic elevation on land or sea floor can be a first indication of a salt dome. On land, these features are hunted out and shot. At sea though, the usual approach is to shoot a grid of say, 2 by 3 or 3 by 4 miles. The salt domes located by the grid are then closely detailed. With a good pattern of detail lines, a good picture of the shape of the dome can be obtained. The picture is fairly accurate in three dimensions.

On a seismic section, a salt dome appears as a "dead" area, with few or no reflections, extending upward from the bottom of the section. Horizons may be bent upward as they approach the dome. Those above it may show drape over it (Fig. 11-2).

Once a good picture of the dome is made though, the problem is not roads if the features sought are much smaller than a square mile in area.

After such lines, detail lines are required anyway, to locate a well. So, instead of looking first in places you know you won't be able to drill, the original search should be conducted with lines through legally-drillable locations.

This is an economic restriction, so of course it does not apply to the same extent to more economically-obtained data. Thus seismic information purchased through data exchanges, traded for, or included in a farmout deal, can form a network on which further shooting is based. However, the additional shooting should not necessarily be only used for detailing leads found in the old data, but can reasonably be used to fill in the spaces between the lines. This can be in stages, of course, first following up the leads. The larger gaps in control, though, should not be considered as unprospective, and should be checked when economic, land, etc. conditions permit.

Some areas of this type have had some wells drilled on them, got some production, but the dry holes have discouraged the industry. From experience in seeing different companies come into such an area, a pattern is evident. A company starts with the large-structure assumptions—that reconnaissance lines should be shot on roads for economy, that fairly wide control is adequate, that a well can be drilled off of the control as long as it is on the contoured structure. The company either drills a few dry holes and leaves the area, or is lucky at first, finds some oil, and starts to learn the lessons of the area. At any point, it can fail to learn, and pull out. So, a number of dry holes drilled by the industry in general do not condemn such an area. They just indicate the companies that used the wrong approach. If, in an area of small features, you hear of a dry hole drilled on a structure, the next questions to ask are whether the structure was detailed sufficiently, whether the highest point was drilled, and whether the drilling location was on a seismic line, or was based on contours a quarter mile or so away from the line.

Salt-Solution Features

In the salt-solution areas of the Williston Basin, there are many small fields found, and, surely, to be found. They require a special outlook and technique. The areas already developed to some extent are around Out-

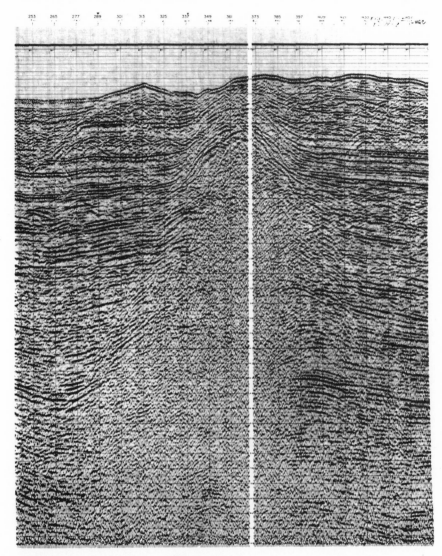

Fig. 11-2 Salt dome section in offshore Lousiana

Credit: Western Geophysical Co.

yet solved (Fig. 11-3). There is still drilling to be recommended, and the shape of the dome and the beds around and over it are not sufficient information to determine the best drilling location. Oil may be trapped in

a bed upturned against the dome,

a pinchout in an upturned bed,

drape over the dome,

a fault trap,

a bed upturned against a shale sheath on the dome,

under an overhang of salt.

So there may be oil on any side of the dome, or directly over it, and on the flanks many depths should be tested. About the best thing to do when you have a dome outlined and leased is to plan a program of about five wells, to which you commit yourself in advance, and decide not to get discouraged until all are drilled. This five is a sort of minimal number. In the Gulf Coast, there is talk of this dome being productive and that one dry. But the "dry" one may just not have yet been drilled at the right location.

Salt, shale, and igneous intrusions can look alike on record sections. Any of them can form oil traps, in the same ways. If there are reflections in or under them, they can be distinguished from each other by velocities derived from the seismic data, as described in "Velocity Analysis", in Chapter 5, Data Processing.

The amplitude anomalies given under "Bright Spot", in Chapter 8,

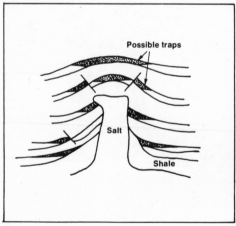

Fig. 11-3 Salt dome traps

Interpretation, are an aid to finding the gas and oil associated with salt domes.

Reefs

Reefs can appear in very different ways on seismic sections. They can be clearly visible (Fig. 11-4), subtle and difficult to delineate, or totally invisible. And these various degrees of detectability can occur near each other, even in adjacent reefs, although more often they are typical of different areas. So the reefs can be detected on seismic sections directly, or indirectly, or not at all.

Seismic reflection takes place at velocity contrasts. The limestone of a reef, if it differs in velocity from the rock over and around it, will produce a seismic reflection. Some reefs are overlain by sand or shale of a good bit lower velocity than the limestone. Or the overlying material may include some carbonate itself, and have a velocity close to that of the reef. A reef containing gas will have a lower velocity than limestone with water in its pores, so may have a velocity similar to that of nearby other rocks.

The greater the contrast, the stronger the reflection, so these various situations can lead to all degrees of reflection strength.

The velocity contrast, if there is enough, makes for a reflection at the contact between carbonate and the other rock. There isn't much way to distinguish between reef in place and reef detritus. Both are carbonate.

The well-defined reefs, for instance some in Indonesia, can be easily seen on sections, and well mapped as to area and upper surface. Interior features, layering, etc., are generally beyond the present ability of seismic exploration. Thickness of the reef can be determined only if the reef is thick enough and there are good enough reflections at both top and bottom, and the reef is shallow enough, to make a good RMS velocity determination of the reef material, so thickness in feet can be calculated.

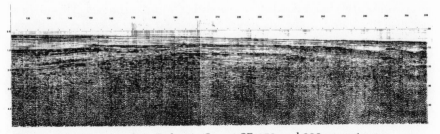

Fig. 11-4 *Apparent reefs in Balearic Sea at SP 150 and 290, near 1 sec.*

Credit: *Digicon Inc.*

The less clear reefs may be, in some cases, the more productive ones, as a reef with its open spaces filled with gas, or even oil, will have a slower velocity of sound than one containing water, and so may not produce as strong a reflection. The subtlety can be such that a reef is fairly clear on one section, and not seen at all on a parallel line, or even on one intersecting the first within the reef area.

When the reef itself does not show up on a section, there may be indirect evidence that will indicate its presence, and even its outline and thickness.

The carbonate reef material is not as compressible as the surrounding formations in most cases, so differential compaction causes overlying formations to be depressed around the reef, but not over the reef. This drape is greatest in layers closest to the reef level, so mapping of a reflection just above the reef level can indicate, by its locally uplifted places, the presence of reefs below it.

In a fairly thoroughly explored area, a numerical estimate of reef height can be obtained from the drape. This is a matter of experience in shooting and then drilling wells into a number of reefs in the area, and is only a rule of thumb. Some number of milliseconds of drape will represent so many feet of reef. This relationship won't usually stay constant over a large area, so has to be worked out for the parts of the area.

In an area of fairly flat horizons, the interval maps are valid in themselves. But, looking for reefs in a more tectonically disturbed area, a structure map may also be necessary to determine the tilt of the reef, so a well can be drilled at the highest part. Spilling out of the oil from the upper end of a tilted reef may be prevented only if the material underlying the reef, as well as that above, is impermeable.

Also depending on the surrounding rock, the velocity of sound in reef material may be faster, slower, or the same as in the surrounding rock. In most areas, reef velocities are higher. This causes the phenomenon called velocity pull-up. A reflection from a layer below the reef may show some apparent raised structure directly under the reef. This occurs because the sound, traveling down to the layer and back, accomplishes the trip more quickly where its path is through the reef, than at other places. The pull-up is not real structure, but is useful as another clue to the reef. And, of course, an apparent structure under a reef does not have to be pull-up, but could be an uplifted part of the old seabed, on which the reef started to grow.

When the surrounding rock has the same velocity as the reef, then there will be no velocity anomaly. Also, when the velocities are the same, the compressibilities are about the same too, so there may not be any drape in such an area. This makes reefs difficult or impossible to find with present-day seismic exploration—naturally, as velocity differences are what seismograph depends on to yield any data at all.

In areas where the reef velocity is slower than normal for the depth of the reef in that area, the pull-up phenomenon will occur in reverse. A horizon below the reef level will seem to be depressed under the reef. This pull-down is just as good a clue as pull-up.

Reef indications tend to vary from place to place. So it is best to have some experience in the particular area. If it's a fairly new area, but with some oil already discovered, then it is very important to get at least a line of seismic data over the area, by trade or purchase, or by shooting it. Failing this, well data can help somewhat in deciding how a reef might appear on a seismic section.

In some areas, a reflection may be interrupted by reefs poking through it. So the absence of the reflection may be a reef indication.

A thorough velocity analysis of a section may give additional information on velocity differences between reef and non-reef. Velocity data comes from reflections, so this depends on a strong reflection at the top and one at the bottom of the reef, or at least, one above and one below the reef. However, as reefs are not usually very thick, say less than 1000 feet thick, there is not much effect on the overall velocities to the horizons, so the velocity data may not give very solid information.

Geological processes related to the reef, either as it grew or resulting from its presence, may influence velocities above it. Differential compaction is one such process. There can also be differential water movement, cementation, etc. So velocity investigations of zones close above the reef may also help detect it.

These same processes may make layering thinner, or anyway different, over the reef, so analyses of seismic frequencies may be useful. They are made by frequency analyses, as used in filter tests.

Reflection character, influenced as it is by thickness changes and lithologic changes, may be different in horizons above reefs. So careful examinations of character may help locate otherwise elusive reefs.

Pinnacle reefs are small in area, so have the special requirements of small features. They call for closely spaced seismic lines, and wells drilled on lines and at high points.

Stratigraphic Traps

Seismograph has conventionally been used for finding structures. It indicates direction of dip better than any other parameter, and with this and time to a reflecting horizon, does very well at finding structures. Stratigraphic traps, though, are generally found only by drilling. The thought that seismograph might be adapted to finding strat traps, and do so as well as it finds structure, is a tantalizing idea. Much effort has gone into adapting it to this purpose, with a fair amount of initial success.

In some borderline cases, seismic exploration has already been finding oil quite well. These are usually situations in which a non-structural trap has some structure or structure-like feature associated with it.

Reefs are found by mapping the upper surface of the reef or the drape in an overlying bed. The reef top or drape is similar in seismic appearance to structure, and can be mapped by picking a horizon on the sections.

Pinchouts formed by differing dips above and below an unconformity are detectable seismically. So, sometimes, are those formed by depositional termination of a bed. In both cases, it is a matter of picking reflections. But, also in both cases, the reflections do not continue reliably all the way to the pinchout. The beds can be mapped as they approach each other, and then the reflections blend and change character, so they can no longer be followed with confidence. The two horizons get so close together that a different combination of wavelets appears on the section. Judicious projection is then necessary to determine where the actual pinchout is. Sometimes synthetic seismograms, which were discussed in Chapter 7, Data Handling, may help locate the point of pinchout.

Lithologic Variation

Lithologic differences form traps that are nearly "pure" stratigraphic, from a seismic viewpoint. If a formation is porous downdip, and becomes impermeable updip, it can hold oil, but all that structure contributes to the picture is the regional dip, no different from other regional dip.

So normal seismic mapping of a horizon will not determine that the trap is there, but much thought and effort has gone into detecting such traps, and in some special cases they can be found.

A reflection character, as seen on a section, is not a single wavelet, but is a complex of reflections including those from any minor layering nearby. If there is a change of rock type, or of thickness, the character of the complex will change. This change of character is brought about by the changes in strength of the wavelets because of velocity changes from the different types of rock, and different combinations of wavelets as thicknesses change.

If the characters could be identified, and classified into groups like a limestone character, a sandy character, etc., that would be great. But they can't. Seismic sections don't give that kind of information at present.

What can be done is that the characters of a reflection can be calibrated in an area that has lots of wells, and seismic lines through the wells. Then it may be that, for *that* reflection, in *that* area, the character changes correlate with the lithologic or thickness changes in the wells. If so, then it can be possible to predict a rock type, or bed thickness, before a well is drilled.

A synthetic seismogram can be used to make similar predictions, by modifying a sonic log and making a synthetic of that changed log— changed by, say, cutting out part of the thickness of a formation.

RMS velocities from seismic data can, to a limited extent, provide information on lithology. Especially if a formation is thick and shallow, the velocity data may give clues to its makeup.

And, the greatest hope of all is in the direct determination of hydrocarbons, regardless of how they may have been trapped. A good bright spot may indicate gas—whether trapped structurally or stratigraphically. With improved definition, it may soon do the same, reliably, for oil, and give quantitative information as well.

12

General Considerations

Participation And Proprietary

Two variations on the ownership of seismic data have become fairly popular.

The simpler is the proprietary survey, or speculation shooting. A contract company shoots an area at its own expense, and then sells copies of the record sections, tapes, etc. to as many companies as will buy them. The survey is offered as a unit, or in smaller portions (Fig. 12-1).

The participation survey does not take as much money to start, but pays for it in complexity. A contractor advances a plan for a program of shooting, and presents it to the industry. Companies choosing to participate each pay a portion of the cost. This amounts approximately to their dividing the cost among the group. The participating companies have a say in the plan, so it usually is modified after being subscribed. The contracting company that proposed it, conducts the survey, and gives copies of the results to the participants.

After this initial phase, the data can also be sold to other companies. There is usually a provision for some of this money to be distributed among the original subscribers, or to be used to obtain additional data for those subscribers. At some later date, when sales are presumably less likely, all proceeds from future sales may go to the originating contract company.

The effort involved in several companies' cooperative decisions is considerable, but those companies do wind up with data that better fits their needs.

Survey Problems

Surveying—just knowing the locations of points—is a never-ending source of problems in geophysics. It is usually taken for granted, and accepted as correct in the interpretation stage. The "accuracy" of a transit survey, or shoran, or satellite, is well known. It may be given as so many feet in a certain distance. But this is its accuracy, *when* everything goes right.

On land, there are checks that can be made after a line has been surveyed. Metal tags on bushes, piles of dirt by old shot holes, cleared paths in the jungle, can be found later, and re-surveyed if necessary.

At sea, there are no marks left after a line has been shot, so the line can't be re-surveyed. It can only be re-shot, that is, a new line shot at what is supposed to be the same place. There is one kind of check that can sometimes show that there is an error. Where two seismic lines cross, if a clearly recognizable reflection appears much deeper on one line than on the other, then *something is wrong*. That can't really be the intersection, or the reflections would be at the same depth. Or a prominent feature—fault, reef, salt dome etc. may form a coherent picture on some lines, and maybe another coherent picture on another set, but out of accord with the first. If shifting one set of data makes the picture fit together a lot better, then probably one set is in the wrong location. Which set? The misfit just indicates the problem, but doesn't give any clues to which set should be moved. You can tell only if one set of data is firmly tied down by wells, water depth data, ties to landmarks like buoys or islands, etc. And the tie to well data must be considered suspect. Seismic data and

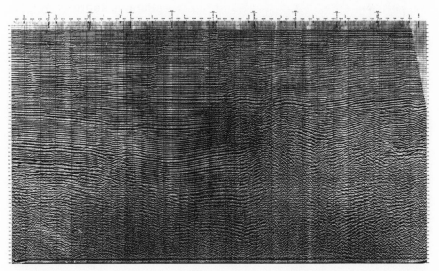

Fig. 12-1 Proprietary section, in offshore Texas   Credit: Teledyne Exploration

well data are too easy to rationalize into an apparent fit. Only if there is a definite mis-tie can you know. And then you only know something is wrong.

Probably the best way to handle surveying is just—*very* carefully. That is, whatever survey system you use, give it a good bit of attention—don't take it for granted. Talk to the surveyors, get them to explain the problems, spend the money for extra equipment, extra checks, etc., that they recommend. Above all, pay attention to the surveying.

Seismic Reliability

Seismic exploration is expensive, difficult, requires much time, needs the efforts of specialists to interpret it, and then maybe results in a dry hole. Why is that?

There are a few sayings to discuss in that connection. "The seismic was wrong, we hit the Devonian 50 feet low." "But we need a location over here for the land situation." "If it won't produce here, it isn't big enough anyway."

The sayings all have to do with a philosophy of seismograph, and what it will, or should, do, and what it won't do. This drilling wells business is expensive, and you don't want to drill dry holes. Even more, you don't want to miss a field, decide it doesn't exist, and leave it for somebody else to find. To reduce these risks, exploration people all need to speak a common language, so oil fields aren't missed just through a communication problem.

All that sounds kind of to one side of the subject, but does indeed tie in with the use of seismograph.

There are several things you would like to get from a scientific, expensive method like this. Some of the things it does better than others, and some not well at all.

The one thing seismograph does best is to give direction of dip along the line shot. That is, if it says a horizon dips to the right-hand end of a line, it probably does. There are some situations, like rapid velocity changes, that can make even this indication wrong, but it's the thing the seismic method is generally the best at.

In some areas, this is all that is needed. I worked an area in Montana in which I couldn't tie wells, but could determine a point that was higher than nearby points around it. These local bumps enabled us to find oil much more consistently than we would have if I'd devoted my time to making the data tie the wells.

Amount of dip isn't nearly as well determined by seismograph as is direction of dip. Velocity enters into the picture in trying to determine how much dip there is, and makes it less reliable.

Absolute depth would be a very nice quantity to obtain, but here again velocity is a big factor. Well tops might be picked wrong, but if they are on the right log kick, they can be read down to the foot. Seismic data aren't like that. First, the wiggles on a trace are long loopy things compared with well log curves. One seismic peak extends over about 150 feet. So, in general, information can be obtained only about every 150 feet in the section. That's a bit different from fine wiggles that respond to change every foot. Also, only times are read from the records, and they have to be converted to depth with velocities that are approximations.

So we can miss in predicting depths to formations. I've frequently had to explain being wrong by 200 feet, and once, in the Colorado Rockies, 900 feet. In the latter case, all I could do was guess there must be an undetected fault between the two wells that I didn't tie.

Another thing that should be pointed out is that seismograph can't predict situations where it was not shot. This sounds simple and obvious, but a location is often selected from a contoured map, without much regard for actual data. A large feature, of course, can be seen to build up over a considerable area, and have a wide latitude for drilling on the high part. But, now that many areas are being explored for little bumps, reefs, salt collapse structures, and the like, this big-feature philosophy won't work everywhere. A well, in one of these areas, should be drilled on a seismic line, and at the highest point on the feature. A location a quarter of a mile away could just be off the feature. So a dry hole would be drilled presumably on a basis of the seismic data, but actually not testing it at all. Contours are just guesses where they are not on seismic lines.

Then there is the "most tolerant position" philosophy, which says that, when several kinds of information are available, the best position is one that looks good on all of them, not the one that looks excellent on just one of them. Kind of modifying this to fit seismic data alone, it would say that the best location should be drilled, not one that looks marginal. The "If it doesn't produce here, it isn't big enough" idea assumes that the mapping is either perfect or worthless. Suppose there is a structure on a seismic map, large enough to contain several well locations. There are some reasons to drill one of the lower ones first, rather than the highest position. But maybe the mapping is correct in that there is a structure, but doesn't have it outlined exactly right. Then the only way to be pretty sure of being on the structure at all is to drill the highest point. If that produces, development can proceed. But if a low position had been drilled first and found to be dry, the whole prospect would have been abandoned. The high point may not turn out to be the highest, but it is the most likely to be somewhere on the structure. And the structure may even be as large as indicated, but not shaped just as mapped.

So seismograph will not do all the things that would be nice to have done, but the thing it does best, pointing directions of dip and therefore, locating locally high places, is very useful to finding oil.

Be glad it isn't the other way around, a method that ties wells perfectly, but can't isolate structures. Where the wells are, you already know about.

Geological-Geophysical Coordination

There has been much talk about collaboration between geologists and geophysicists. The collaboration is necessary in order to keep the two groups from drifting apart, too proud of their own specialities, and unwilling to concede much to each other. Management wants coordination. Geologists want it. Geophysicists want it. Some of them, anyway.

Yet the talk continues, and the groups continue to be more separate than efficiency calls for. The problem, of course, is that the people concerned are busy, and are trying to get correct answers in their own disciplines, which is hard enough, and just don't get around to working together sufficiently. And when a person does come up with a plausible idea in his own specialty, it isn't fun to have it knocked down by somebody in another field altogether.

A few of my experiences have been pertinent to the problem. In one office, there were a geophysical department and a geological department. The very existence of the two groups, under different bosses, made them operate as separate units. But, our offices were on the same floor of the building, so we occasionally knocked heads on projects. There was plenty of formal get-together, but the productive sessions occurred when somebody wandered into the other department with a cup of coffee in his hand, or when a guy was so excited about a prospect that he had to tell people. What happened on several of these occasions was that we in our department, had a neat picture of the area, built up after rejecting several ideas. This new idea was good, and could be defended easily. The trouble was that the geologists also had a good, solid interpretation. And the two were contradictory. There would be a period when each thought the other must be out of his mind, to miss such obvious evidence. Then we'd argue it out, pitching out what couldn't be, from one viewpoint or the other, and clinging to what had to be. Eventually, a new idea would be hammered out, that might be quite different from either of the original ones, but that fit everything well. All of us came away pleased and pretty self-satisfied.

In another office, a geologist and I were assigned a long-term project, and established in adjoining offices. I'd get some more data, and run into the next room. Or he'd fit it in with some wells, and run into my room. Then we were transferred to offices two thousand miles apart, and retained the assignment. We still worked on the same project, but by long-distance telephone and mail. Continuing the collaboration, we worked well together, but there was something missing.

Other times, a geologist has worked up a prospect, I've shot it, worked up the seismic mapping, and presented it back to him. Then he was supposed to take over and make a recommendation. This often worked well enough, but there was always the chance that one of us would bring up some simple, obvious thing that the other hadn't known about. So a lot of the work would unravel. Worse, the simple things might not happen to get mentioned at all, and the interpretation would remain, based on an oversight.

With all this reminiscence, I don't have a really good way to handle the problem. If there was one, it would be in use universally by now. Much depends on personalities, the company's history, how many people are in each department, whether consultants are used, and other administrative-type factors.

The one contribution I'd like to offer is that groups that work together get an esprit-de-corps, an us-against-the-world feeling. So the geologist and geophysicist working on a project should feel themselves in the same little group. I don't think departments should be reorganized, but maybe a geologist and geophysicist working on the same area could be somehow identified as a unit, and discuss and work together, so they thoroughly understand each other's viewpoints, even though they still may not agree. This will work best when both are well-qualified, experienced people.

This joint operation is probably better even than the assumed ultimate of having one person who is both geologist and geophysicist. The one person has more trouble spreading himself thin enough to work effectively on the two kinds of data, and the interaction of two minds is lost. And he has difficulty in keeping his prejudices and way of thinking in one field from influencing him too strongly in the other. There is real benefit in having two people with different outlooks work together on a problem.

Factors In Program Planning

A general discussion of program planning was given in Chapter 3, Field Operations. But in planning a seismic program, there are also a number of special factors that need to be considered.

First, of course, is whatever the reason for shooting may be. It might be to investigate a prospect found by the use of well data or surface mapping. It may be a matter of covering scattered leases, or a large exclusive permit area.

If an earlier seismic survey is available, it will dictate some features of the new program. Probably, new lines should tie between old ones. Critical old lines shot with a poorer technique than the present ones, may need to be duplicated, to see if the leads are real. Large gaps in the older control might need to be filled in.

If there are wells in the area that are not tied in with seismic control, they probably should be tied. In some areas, though, notably in the Williston Basin, well ties may not be very useful. If ties between wells do not give the correct dip, but shooting does isolate locally-high features that produce oil, then a meticulous tieing of wells and trying to make the ties fit, may divert effort from oil-finding.

More important than just well ties, in a new area, is having some control over a producing field, to show what the things that produce look like on a record section. In fact, any time control over production is available, it can be instructive.

There tends to be some difference in outlook between geophysicists and geologists, in that geophysicists like to shoot loops, to see if interpretations tie back on themselves, while geologists prefer tieing wells, to anchor the interpreted horizons to them. Both are devices for making the interpretation somewhat more sure. Both are useful, but either can be overdone. Tieing loops helps where correlations are doubtful. Tieing wells usually works, but may sometimes misdirect the major effort. Shooting a line with neither loops nor well ties, is often good enough for regional control, or in areas of very reliable data, but has no safeguards, so when it is undertaken there should be an awareness of its limitations.

Terrain and other physical obstacles of course need to be considered. This is obvious, but looking at a nice white map, a person tends to forget that there may be hills, streams, canyons, railroads, towns in the way. A government topographic map helps in planning a survey. Air photos are also good. But whatever aids are used, even scouting the area in advance, the unexpected may crop up when the program gets underway. Permit problems and new constructions, for instance, do not appear on the maps. Therefore, someone on the spot needs to have the authority to modify the program to fit conditions. To do this effectively, he must understand the program and why it was laid out as it was. He also needs to be competent to judge requirements for seismic lines.

Usually, this person in the field does not exist, so the survey is planned as well as it can be, and the field crew shoots it if at all possible. This has led to some rather foolish lines being shot—alongside a recently-cleared line, or across an obstacle that was not known about, or did not exist, when the planning was done.

Often a line is planned in a certain way just so it can be put on paper. That is, if lines have been shot all around an interesting-looking place, and a line through it is needed to make the pattern of control closer, which way should the line be run? North-south? East-west? Diagonally? It may not matter, but something has to be put on paper, so it is drawn, say, east-west, for some minor reason. In the field, it turns out that a new road is being built right through there in a north-south direction. The crew doesn't know there is any latitude in the program, so they shoot across the road work. Some geophones are set on loose, graded earth, giving poor record quality. Maybe a cable is cut by a grader. If they had known, they could as well have shot north-south, parallel to the road work, and out of its way.

There isn't any real way around this problem, except having a competent person who understands, and maybe helped design the program, in the field. Not much latitude can be given to a field crew for two reasons. Anyone down the chain of command can get afraid of the responsibility and jell the program as drawn. And, someone can misunderstand and use more latitude than was intended, maybe repeating a line he didn't know you had, or getting off of the prospect that was to be investigated.

A program needs to be fitted to the type of feature being sought. A large structure can be hunted most economically by shooting lines along roads. A small one, like a pinnacle reef, calls for lines shot through drillable locations. In mountain fronts, dip lines are particularly important.

A seismic program should never be planned without, at least, the participation of a geophysicist. Geology, leasing, and geophysics each have requirements that must be taken into account in planning. Geology or land requirements produce the prospect in the first place, and geology provides direction of regional dip and other characteristics of the prospect. Geophysics dictates requirements like tied loops, close control, multiplicity of stack, etc.

If the program is laid out, assigned to a crew, shot, and then a geophysicist is called on to interpret it, much of his value on the prospect will have been lost. Program planning is as geophysical an activity as interpretation.

Continuing Program

When a company has a large area to explore, for instance a production sharing contract area or concession, there are special situations that indicate the way seismic programs should be conducted to serve their purpose best.

The fact that it is a large area, and is to be explored for a number of years, means that the shooting needs to be adapted to both regional and detail purposes, and that the immediate reasons for a certain shooting program should be balanced with longer term requirements.

At the start, any data that already exists should be obtained. This will help with the regional mapping and is more economical than new shooting.

The first shooting the company undertakes in the area may need to be experimental, to determine what record quality can be obtained in the area, and what seismic techniques will best fit the area. It may also determine how reconnaissance data can be shot, at perhaps less expense than a normal survey.

Early in the exploration, a rectangular grid pattern for seismic lines should be established. It might be a north-south and east-west grid, or lines may be run parallel to a coast and perpendicular to it, or on a grid determined by regional structure.

In later shooting, it is advantageous to try to keep to the grid pattern. This will tend to eliminate the recrossing of prospects in odd directions that so often occurs when a program is allowed to "just grow". The parallel lines will give close control for less money. Criss-crossing lines are close together near the crossing point and farther apart at other places, so it is clumsy and expensive to add more data to them. A grid can be filled in with more parallel lines at any time. As new techniques are used on later lines, they will fill in the grid with additional data, or, if only the new lines show some particular subtlety, they constitute a similar but more open grid themselves.

Some special conditions will call for lines in special directions. These should of course be shot but fairly sparingly.

In particular, as wells are drilled, it is a good procedure to tie them together with lines through the well locations, direct from one to another. These lines should be shot with the very best technique available. When a newer method comes along, they should be re-shot. They are the best testing ground, as the wells serve as a check on the shooting. Solid information between wells will also increase the information yield of the wells. These well to well lines can constitute a backbone for any difficult interpretation situations. Once the interpretation of those lines is resolved, the other lines can be worked outward from them and tied back into them.

Development Of Oil Fields

The number of seismic exploration lines to be shot on a prospect is limited by the fact that the prospect may not produce. So a reasonable amount of shooting will establish the presence of the prospect and a fair idea of its extent.

But, once a field is discovered, two things are going to happen. A lot of money will be spent, and a number of installations will be set up. The money should be spent with good technical information. The installations will hamper or preclude future shooting.

In the flush of success after discovering an oil field, a company tends to feel that the seismic work has been done, and was successful and that the time is now at hand to drill delineation wells, gather production histories, and work toward the development of the field. However, in the following few years, there will be a number of occasions when more information on the shape and other details of the field will be needed. Locations of wells, platforms, and other facilities will have to be made in part by guess in the absence of more data.

With a field partially developed, seismic data is more difficult to obtain, more expensive, and often must be done without.

The time to do a detailed seismic survey of the field is right after it is discovered. Particularly offshore, where the rig may move somewhere else after drilling the first well, there is an excellent opportunity.

At this point, a number of lines at very close spacing should be shot. "Very close" is a matter of the spacing normally needed for exploration in the particular area. The "very close" lines should be, at the very least, twice as close as exploration detail lines. They should extend some distance beyond the edge of the prospect, both because the prospect outline of the moment may not be correct, and to give a good solid idea of the regional characteristics around the field. The lines should be in a consistent plan to yield data that tells all about the field, not specific lines individually planned to answer certain questions.

Tomorrow's questions won't be the same ones as today's. The consistent plan will, in most cases, be a grid, parallel lines with another set of parallel lines, probably at right angles to them. Once set can be in the general direction of dip for the area.

This group of lines will then be available for future use, as when a development well shows some surprising characteristic that starts people off on a new round of investigations about the field.

For an average field, the seismic expenditure on this post-discovery shooting project will be from one to several hundred thousand dollars. Sounds high, but weigh it against the millions or tens of millions to be spent on wells, and, if offshore, on platforms, production installations, etc., all of which can well use more data to guide their planning. If the seismic survey allows a platform to be set in a better location, an unneeded well to not be drilled, or a needed one to be drilled, the financial investment can appear very small.

All this doesn't mean that the seismic answers will be absolute. They won't. The answers aren't absolute in any other aspect of the business either, or any other business for that matter. But they will give you a lot more of a fighting chance.

The shooting project will also not mean that no more shooting will ever be required. New developments in seismic exploration are being made continually. If next year produces a seismic technique that yields some new kind of data, say quantity of oil or something, then it may be worthwhile to do some more shooting. There may be so many buoys and things around that a seismic boat can't get its cable through where you'd most like the information. So you get what you can.

Drill The Top

When a small feature is to be drilled, the well should be on the highest point possible without getting too close to the edge.

In the case of a simple anticline or a reef, the highest point will be located near the center of the feature. There are several reasons for drilling the top, in this case.

1. Production may be a maximum.
2. Drainage should be more complete.
3. If there are surveying errors, the central position will be most likely to be on the structure.
4. If the contouring is not correct, the center will be the safest.

Note that 3 & 4 are always the case. The survey is never perfect, and contours are always guesses.

5. There are very likely to be surprises at the top of the feature.

No. 5 includes porosity development by winnowing over a structure, development of a reef on the high point of a structure, trapping of hydrocarbons here and there above the feature. This argument applies to not only the first well on the feature. If the first well was not drilled on the top, then a subsequent well should be drilled there, notwithstanding careful attic oil calculations by engineers, based on porosities and other conditions in wells on the flanks (Fig. 12-2).

...... THEN, WHEN YOU CROSS MY FENCE DOWN BY THE CREEK, YOU'LL BE 'BOUT ON TOP OF THE STRUCTURE. YOUR RECORDS WILL START GOIN' BAD, LOTS OF HIGH FREQUENCY STUFF. MOST OF THE OTHER FELLOWS GOIN' THRU THERE USE A WA-30 FILTER. THEN WHEN YOU 'BOUT REACH THE HIGHWAY ETC, ETC. -

Fig. 12-2   J.C. Knight

Most Important Factor

The most important thing in a seismic search for oil in a specific area is—to determine what the most important factor is, and direct the effort primarily toward it.

Sounds obvious. I know. But it isn't done all that much. People tend to have pet things they consider vital, and devote their efforts to, like doing everything possible to get the very best record quality. But maybe record quality isn't the main factor in the area.

In some areas, record quality is the main thing. It's hard to get good records, and attention must be paid to anything that can improve them. Maybe number of lines should be reduced, so more money and effort can be spent on getting a few good lines.

In other areas, good records come easy. An improvement would be nice, but wouldn't find much oil. Maybe money can be better spent on more lines, with less processing.

In areas of small features, the most vital thing may be close spacing of lines, carefully positioned to cross locations that can be drilled.

In areas of complex folding, direction of lines—directly down dip—and best possible migration of a number of lines, may make the difference between understanding the geology and not understanding it.

In highly faulted areas, two things may be necessary, best possible record quality to define the faults, and some sort of three dimensional control to determine the directions of the faults.

In areas of many obstacles to shooting, no-permit patches, rugged terrain, islands, platforms, the main factor may be designing a program to get the needed control by going around or shooting under the obstacles.

In areas of difficult access, logistics may be most important—getting supplies in, having spare parts available, etc.

Once, working for a company that seemed to be putting its main stress on number of profiles (shot points), a friend of mine commented that he thought seismic success was measured not in profiles, but in barrels.

How To Find Oil

This heading turned you on? Thought it might. And I do indeed intend to tell you how to find oil seismically.

1. First step. Have an area with good chances for oil.

2. Have the area worked on by a geophysicist, in all stages, planning through interpretation. He needs to work on it long enough to get his feet on the ground, say a year or two as a minimum. The geophysicist should be competent, imaginative, optimistic. If he is given his head, and not directed too much, there should be surprises.

3. Have plenty of opportunity for interaction between geophysical and geological personnel.

4. Determine what is the most important factor, or factors. This isn't necessarily an early step. You just find it out as soon as you can. It may only take some conversations with people who know the ropes in the area. Or, much more likely, it may take several years of work.

5. If there are oil fields in the area, get some seismic control—old or new—over them.

6. Shoot a program or programs that fit the area. Don't just lay them out. Get seismic expertise in the act.

7. Don't "science it to death". If you get some prospects to drill, think things out thoroughly, raise objections, but don't go through lengthy re-investigations too long. The prospect can't be proved until you drill, so don't expect it to be.

8. Drill some wells.

No guarantees, but this is the best way I know to give yourself a fighting chance to find oil fields.

Appendix

Other Sources of Information

This section is given, rather than a list-type bibliography. It seems preferable to have a discussion of the references, describing what can be found in them. The references are listed more or less in order of usefulness to the person who does not want to, or is not ready to, go deeply into the technical side of geophysics, or the person who knows the technical side, and wants to learn the workaday techniques.

ENCYCLOPEDIC DICTIONARY OF EXPLORATION GEOPHYSICS
(R.E. Sheriff, 1973, The Society of Exploration Geophysicists, Tulsa)

This book is essential to anyone dealing with geophysics. No matter what your activity, there will be unfamiliar terms in geophysics or peripheral fields. It explains clearly. Its main use is as a reference.—"What does _____ mean?" But it is also very worthwhile to thumb through it. You are almost certain to find a chart or definition that you will profit from.

DICTIONARY OF GEOLOGICAL TERMS
(American Geological Institute, 1962, Doubleday & Company, Inc. Garden City, New York)

This dictionary, while primarily geological, defines some geophysical terms. It does so from a somewhat different approach from that used in this book or the geophysical dictionary above. So it is worthwhile to have available also.

COMPANY HANDOUTS

This is the best, and in some respects may be the only, source of up to date information on geophysics. The handouts are mostly from geophysical contractors and suppliers. They are in the forms of advertising brochures, explanatory folders, copies of papers written by their people. The papers may be mathematical, but the other things tell about whatever the company is proud of, and try to keep it understandable. And they can be on subjects not normally published in magazines or books—a new record section display method, new cable, vehicle, etc. These handouts are most readily available at geophysical and related conventions, where many companies have booths in one concentrated area. They can also be obtained from company offices but you may not be aware of what to ask for. It's easier to look around at a convention.

ARTICLES IN GENERAL OIL MAGAZINES

The oil magazines that are not primarily devoted to geophysics—The Oil and Gas Journal, World Oil, Offshore, etc.—quite often have articles on geophysical topics. These articles tend to be practical and understandable to non-geophysicists. Many of the articles are well worth looking out for, and either keeping in a file, or listing in an index, so they can be referred to when needed.

BOOKS

Several books on geophysical exploration have been published through the years. They are good texts for going into the theory of geophysics, and cover instrumentation and field operations as well. The older ones are very out of date, but new books are now coming out.

Particularly useful is Dobrin's "Introduction to Geophysical Prospecting, Third Edition" McGraw-Hill, 1976. It covers about the same ground as this book, but from a somewhat different viewpoint, so would make an excellent addition to what is read in this one. It has more mathematics, but that, as suggested in Dobrin's preface, can be skipped. The chapter on interpretation, Chapter 8, is especially practical, with good explanations and examples.

AAPG BULLETIN
(American Association of Petroleum Geologists, Tulsa, monthly)

The AAPG Bulletin is, of course, devoted to articles of interest to geologists, not geophysicists. However, its articles on particular areas are often illustrated with seismic sections and may tell of seismic problems in exploring the areas. These articles are quite useful if you are planning to shoot one of the areas, or a similar area somewhere else. These articles too could be advantageously indexed or filed for reference.

GEOPHYSICS
(Society of Exploration Geophysicists, Tulsa, six issues per year)

Geophysics is a magazine intended for geophysicists. However, as it worked out, it is primarily for, and by, research geophysicists, so it is mostly pretty technical. Efforts have been made through the years to make it more a magazine for practising geophysicists. For a while there was even an annual issue titled "Papers for the Field Geophysicist." It didn't last, as not enough people submit papers of general interest.

The advertising in Geophysics is something of a short version of the company handouts, and is useful in the same way they are. A "new products" section serves a similar purpose.

GEOPHYSICAL PROSPECTING
(European Association of Exploration Geophysicists, The Hague, Netherlands, quarterly)

This magazine is very like Geophysics.

JOURNAL OF GEOPHYSICAL RESEARCH
(American Geophysical Union, Washington, D.C., three issues per month)

This magazine is a scientist's publication. The "Geophysical" in its title is geophysical in its broadest sense, including weather, volcanology, etc. Occasionally it has an article that is somewhat useful in exploration geophysics.

Educational Requirements

There is a confusion in the minds of some management and personnel people on the education required for geophysical work. The confusion is in part caused by the many aspects of geophysics with their many different requirements, and in part by the general hiring problem faced by personnel departments.

The personnel hiring problem is the same as the one in hiring secretaries. If you ask the personnel department to find a secretary for you, their first question is: "How many words per minute?" For many secretarial positions, the question is almost useless. You want a secretary to keep the files straight, find things when you need them, remind you of appointments, type a few letters, and occasionally type a long report. The only time typing speed makes any difference is in the report, and even then it isn't nearly as important as correct typing, good proofreading, a knowledge of spelling, and good arrangement on a paper. So why do the personnel people ask about typing speed? Because they can measure it. It's a thing people can be graded on, and helps excuse the failure to judge other factors. The failure isn't their fault, as the factors are too hard to measure, but they'll get blamed anyway.

In seismic occupations, the comparable question is: "What kind of degree do you need?" Again something that is easy to determine. Without thinking much, a person may answer, "A Master's in geophysics, unless we can get a Ph.D." This not only isn't the best answer, it may be seriously wrong.

The different aspects of geophysics with, more or less, their different requirements are:

Field Work The main ability here is to be able to make equipment work somehow when things break down. A field man needs to be able to improvise. It helps if he's been making equipment work with baling wire since he was a kid. The usual way to do that is to grow up on a farm. Maybe an adolescence spent fooling with old cars might be a second best. If you can get people like this, you are way ahead. This applies while they are in helper positions in the field, and also later, when they become field supervisors, heads of departments, presidents of companies.

Recording Recording equipment is electronic. Here, a radio ham may be a good type, and/or a person with an electronics degree.

Processing This is a computer field. Education in all aspects of computers and in applicable mathematics is useful.

Interpreting Here, for the first time in this list, is a field that calls for education directly in geophysics. A degree in geophysics is useful. Some courses in geology would help.

In all these parts of geophysics, I have listed the background that should be looked for when hiring a person. Note that a certain degree level was nowhere mentioned. If a company selects for people of high degree level, it is selecting in part for people who like to go to school. The degree should be considered secondary to other factors.

In the December 19, 1975 issue of Geophysics, there was a report on a meeting between professors and businessmen on what each group required from the other in order to improve the quality of geophysical education. It is noteworthy that one suggestion *from the professors* was that maybe a degree wasn't necessary.

Now we'll turn the thing around. Instead of a company's viewpoint in hiring, consider an individual's position. When a person is debating whether to take steps to learn more about the field he would like to enter, or finds himself in, the answer should almost always be that he should. Certainly, a person in any of the parts of the geophysical business will profit from a course in geophysics. The same goes for a course in geology, petroleum engineering, electronics, computer programming, etc. You will do better if you know your specialty better, *and* if you know the other specialties in your field, *and* if you know related fields, outside of yours. And, while working in a rapidly changing field, it is essential to put forth efforts to keep up with the new developments.

In other words, acquiring education is great, but *selecting* for formal education is hazardous.

What Is A Geophysicist?

I was sitting at my desk minding my own business, doing whatever a geophysicist does to earn his pay, when I heard voices in the hall. Apparently a secretary or draftsman was showing a new employee around the oil company offices. She must have just arrived at our floor and pointed to our office.

"That's the geophysical department."
Pause, while she apparently decided a little more explanation was in order:

"They're geologists."
Another pause. Not quite right yet.

"Special kind of geologist."
Pause.

"The real geologists are upstairs."

For Non-Oil People

This book was aimed at a certain audience, one that has reason to learn something—about as much as one book will hold—about seismic exploration. But, other people, who don't have this incentive, may ask what it's all about. This section is for you to show them. It's the explanation a geophysicist might give a farmer, in the process of getting a permit, a farmer who was interested in what was going on, beyond the usual "We want to drill some test holes." Or to a relative who, to the geophysicist's delight and amazement, shows an interest in what he does for a living.

So, on the following pages, is a brief explanation to show people who are somewhat interested, but shouldn't be exposed to "more than they wanted to know about geophysics."

And, if you are a geologist, production man, or other specialist, pardon me for getting into your field. But, this is for people who don't know about the oil business.

How We Find Oil By Seismic Exploration

Oil, petroleum, crude oil, whichever name you like, is found underground, usually very deep underground. It doesn't occur in a stream, pool, or lake of liquid, runny oil.

Oil occurs in rocks, that is, in any open spaces there are in the rocks—between the sand grains of a sandstone, in the cracks in a shattered rock, in the little cavities in a limestone. The grains of a sandstone are packed very tightly together. Buildings are built of sandstone, and don't seem to have any open space in the blocks. But the sand grains are mostly round, so there are tiny spaces. A sandstone building may sop up rain, and stay wet in spots for days, as water seeps out of it.

Fig.11 Little chunk of rock, with water or oil (black) between grains

Much of the earth is covered with sedimentary basins, places that once were seas. Through the years, dirt, sand, etc. were washed from the land into the seas, forming layers of sediment, stuff that settled through the water. As more layers were added, the weight of them compressed the earlier layers, turning them from mud, sand, etc. into rock. The gases and liquids were squeezed out of it except in the remaining pore spaces left by the shapes of grains. The fluid left in the pores was mostly salt water, from the old sea.

Fig. 2 Old sea in cross section, with sediment under the water

The sedimentary basins may, since they were formed, have been warped, deformed, distorted by various forces in the earth—volcanoes, faulting (which we feel as earthquakes), salt flow, mountain uplift, etc. The basins may now be dry land, or they may still, or even again, be undersea.

Fig. 3 No longer sea—beds distorted

With the rocks wet with salt water, there may have been impurities in the salt water like bits of plant and animal matter. With time, pressure, heat the organic matter can become petroleum and natural gas, scattered through the wet rock. But oil floats on water, so, if the rock layer is tilted, the oil, through the ages, will creep upward. It will

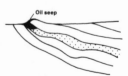

Fig. 4 Detail of Fig. 3

work upward clear to the top of the ground and be an oil seep, if its route is uninterrupted.

As a result of the deformations, though, there may be some local high spots, where the rock is pushed up. If oil floats to one of them, it can't get out. It is trapped. This is an oil trap.

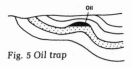

Fig. 5 Oil trap

Gas is lighter than oil, so any natural gas in the rock will float to the top of the oil.

Seismic exploration is a means of looking for oil and gas traps.

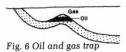

Fig. 6 Oil and gas trap

If you make a loud noise, the sound may echo back from a cliffside or building in the vicinity. If you count seconds to see how long a time it took for the sound to return to you, you can figure how far away the cliff is. Sound travels about 1,000 feet a second in air, so multiply 1,000 by the number of seconds you counted. That shows how far the sound traveled. If it took four seconds then it went 4,000 feet, in going there and back. So the distance is half that far, 2,000 feet.

Fig. 7 Echo

That's what we do with seismograph, but in more detail. We make a loud noise, usually with some kind of explosion, and listen with very sensitive instruments for echoes. We aren't interested in learning how far away a cliff or high-rise building may be, but do care how far it is to a rock layer under our feet. We try to make most of the noise at a pitch that penetrates rock well, which happens to be lower tones than some people can hear.

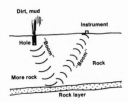

Fig. 8 Underground echo

So, all we do is go someplace, usually a place selected by a geologist who has worked with information from wells, etc. Then we make a noise, have instruments detect it and record it for later reference, go someplace else, and keep repeating. Finding how deep some rock layer is, from place to place will show where it is pushed up, higher than normal. This pushed up place may be an oil trap, so it may be a good place to drill for oil. We can't tell for sure, but the odds are lots better than drilling just anywhere.

Briefly, in practice—

A plan of places to be investigated is made up,

Surveying is used to get to those places,

Explosions are made a little way underground, on the ground, or underwater,

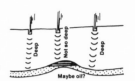

The echoes of those explosions are recorded,

Fig. 9 Three underground echoes—not at same time

The recordings are made into cross sections of the earth to a depth of a mile or so,

These cross sections are used to predict where oil is likely to be.

We do seismic work on land and at sea. Those sedimentary basins can be either place.

On land, it goes like this, but with variations to fit local conditions.

A seismograph crew is assigned a certain area to "shoot" (do seismic work in).

A permit man visits the owners of the land, and asks permission to shoot.

A surveyor with an instrument on a tripod sights through the instrument to a painted board held upright by a rodman. They do this here and there to survey the locations to be shot. It won't do any good to know which points showed the rock layer to be pushed up, if we don't know where those points are.

Fig. 10 Surveying

A drill, like a junior version of an oil well drill, usually built onto a truck, drills holes in the ground and dynamite is put in the holes. The holes are about 60 to 100 feet deep, sometimes as shallow as 20 feet or as deep as 300 feet. The dynamite is loaded in holes to impart explosive force into the ground, where we want it, rather than into the air where we don't, and where it would be dangerous. There may be a small tank truck to supply water for the drill, for bringing ground-up rock out of the hole.

Fig. 11 Drilling

People lay out a long cable from a truck, and by hand, plant little geophones (fist-size or smaller) on the ground to detect the sound; and connect them to the cable.

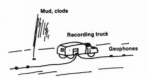

Fig. 12 Shooting and recording

A recording truck, full of electronic sound-recording gear, is connected to the cable, and through it, to the geophones. A cry of "steady" warns everybody to not move for the next few seconds. Footsteps would be recorded. The dynamite in the hole is detonated. Muddy water may shoot out of the hole, maybe 20 feet into the air. The sound and its echoes are recorded for about six seconds on magnetic tape, like in a home tape recorder.

Fig. 13 Reel of magnetic tape

The hole is plugged and filled in, and then the next hole down the line is shot.

The holes, called shot points, are usually about 200 feet apart in long straight lines, and there is considerable running back and forth to keep the cable and geophones advancing to fit the new shot points.

Permit man, surveyor, drill, recorders may arrive on different days. The lines to be shot are planned in advance and the various parts of the crew try to arrange their work so no part has to wait on another.

There are variations. Instead of shooting dynamite in holes, the sound may be made by a truck that presses a plate against the ground and vibrates. Or a propane and air mixture may be fired in a chamber under a truck. A heavy weight may be dropped. In these alternatives to dynamite, the sound isn't loud enough, but a number of sounds are later added together on magnetic tape.

The magnetic tape from the field is taken to a computer center. A lot of sophisticated processing is done, to make the data recorded be more useful. Recordings of lots of shots are put together to make a record section—a sort of cross section of the earth on paper.

An interpreter sits down with the record sections—one for each straight line that was shot. He picks out the echoes from a rock layer and determines how deep it is at a number of points. This information is put in the form of a map, contoured to show the high and low places for the rock layer. If he finds a good high place that might be an oil trap, he may suggest that the oil company drill a well, to see if there is oil there.

Fig. 14 Record section

After all this, about 9 times out of 10, there isn't. So the would-be oil well is plugged and abandoned. It is called a dry hole, or even a duster, although it isn't dry, or dusty. There just isn't oil in it. But, remember that salt water? If there was a market for salt water from ancient seas, the drillers of "dry" holes would be happy.

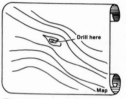

Fig. 15 Contoured map

And while we're at it, in spite of the insistence of television and movies, a well that discovers oil isn't a "gusher" unless somebody has been pretty inept. Blowout preventers hold the oil back. In the old days, gushers wasted much oil, and were horrible fire hazards. The most spectacular sight nowadays is in testing, when oil is allowed, under control, to squirt sideways into a mud pit. Any gas with it is burned as a safety measure. So a good well may for a few minutes have a dramatic 40-foot flame shooting out from it.

Back to seismic exploration. A land crew was described. Offshore the crew is quite different. It is all together on one big boat, 100 to 200 ft long. The seismic boats are seaworthy enough to cross oceans, and to stay at sea, shooting, for a month or more without needing additional supplies.

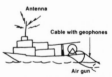

Fig. 16 Seismic boat

The surveying is done, not by sighting through an instrument, but by radio. The radio surveying may be accomplished by measuring the times for radio signals to travel from two special broadcasting stations on land to the boat. This gives the distance from each station, so we know the location of the boat.

Or, the radio surveying may use Earth satellites, receiving the satellite's broadcast beeps, and, from the precisely known orbit of the satellite, a computer on board determines the boat's location.

In either case, one or two guys sit in a cabin with a lot of radio equipment, keeping it working.

There isn't any drilling of holes to shoot in, because at sea, the sound can be made in the water.

Instead of a truck laying out cable and people setting out geophones, the boat tows a cable more than a mile along, with the geophones built into it. There is a huge reel on the boat, to wind the cable on, when not in use.

For shooting, the boat tows a number of air guns. The air guns are chambers that hold compressed gas from compressors on deck, and release it suddenly under water. The pops of the released air are the sound that is then reflected from rock layers far below the sea bottom, and received by geophones in the cable.

In a large cabin are instruments for recording the information from the geophones. There is lots more room here than on a truck. The pops and their reflections are recorded on magnetic tape.

There are skipper, boat crew, cooks, food, bunks, etc. on the boat. Seismic work on land isn't practical at night, but offshore it goes on 24 hours a day. The boat rolls and pitches 24 hours a day, too.

Then, just like for the land work, the tapes are processed in a computer center somewhere, making record sections. Then the record sections are interpreted to find places where there may be oil.

When a well is to be drilled offshore, a rig is floated to the location. The rig is held in place by anchors, or stands on long legs. If oil is discovered, a permanent platform is built on piling, and equipment to handle the oil is placed on the platform.

Fig. 17 Fishing

By the way, that's the place for deep-sea fishermen. Barnacles and things grow on the piling, little fish come to eat them and hide behind the piling, bigger fish come to eat them. The fishing in the area improves dramatically.

Both on land and at sea, if a well discovers oil, then the oil field must be developed. Other wells must be drilled around it to the limits of that field, i.e., until dry holes are drilled. We don't know what the limits are till the dry holes are drilled.

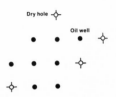

Fig. 18 Developed oil field

Index

Acceleration-canceling hydrophone: on marine cable, 75

Accuracy: compared with correctness, 206-8

AGC: see Automatic volume control

Air drill, 58

Air gun: tuned array, 77; marine seismic source, 77-8

Alias: apparent different frequency, 87; alias (anti-alias) filter, 87, 107

Alidade: surveying instrument, 54

Ammonia print: reproduction of section, 134

Amplitude: in Fourier analysis, 124; shown as color, 135; in synthetic sonic log, 147; amplitude anomaly or bright spot, 208

Analog: form of recording data, compared with digital, 86-7; analog tape, 229

Anchor: to hold explosive down shot hole, 59

Angle of faulting, 189

Angle: of incidence, 6; of reflection, 6

Antenna: for loran-C, 74

Anticline: migration of, 153-4

Apparent dip: of fault plane, 189

Arctic, 62

Assignment map: for field crew, 53

Auto-correlation: in data processing, 116-7; see also Correlation

Automatic gain control, 90

Automatic statics: computer correction of near-surface, 99

Automatic volume control: makes wiggles uniform, 90; also AGC, AVC, automatic gain control, 90

Availability of old data: 224-5

AVC: see Automatic volume control

Band of energy, 179

Bandpass filter, 106

Base line: between shoran stations, 72

Basement: measured by magnetometer, 85

Base station: for shoran, 71; for loran-C, 74; establish by satellite, 74

Bid: for field work, 53

Binary gain: gain recording, 90-1

Binary: number system, 87-9

Bit: on magnetic tape, 129

Boat: for offshore seismic work, 69-70

Breast reel: for portable cable, 65

Bright spot: modeling of, 149; also called hot spot, 208; strong amplitude on section, 208-211; in mapping, 217; at salt dome, 238; in stratigraphic trap, 243

Broad band: recording for synthetic log, 147

Bubble: in marine shooting, 43-5, 80-1

Buried focus: effect of sharp syncline, 154-5; confused with anticline, 214-5

Cable: to convey information from geophones, 4, 65-6; also called instrument cable, 4; streamer, 4, 76; length of, 12, 15; marine cable, 76-7; length for penetration, 109

Cable squirter: to lay out cable, 65

Calculator, pocket: used to make pre-plots, 71

Calibrated sonic log, 142

Canada, 179

271

CDP: see Common depth point
Character: appearance of a reflection, 179-80; changes of, 181; on reef, 241; in lithologic change, 242
Client: representative on contract crew, 54; part of in processing, 129
Coefficient of reflection: used in modeling, 149
Color: sections displayed in, 135
Colorado, 247
Common depth point: also called CDP, 14; shooting method, 14-18; also called roll-along, 66; CDP gather, 102-3; CDP stack, 104
Compaction: effect on velocity, 48
Company crew, 52
Compressed section: horizontally compressed, 137-8
Computer contouring, 204-6
Continuity: of reflections, 176-80; across faults, 185
Continuous line: of shot points, 4, 13
Continuous reflection: in interpretation, 216
Continuous velocity log: well survey with check shots, 142; also called calibrated sonic log or CVL, 142
Contouring: computer contouring of synthetic sonic logs, 147; general, 191-2; geophysical compared with geological, 192-204; by computer, 204-6
Contractor, 53
Convolution: combining of data, 117-22
Corrections: for near-surface, 94-8; static, 98-100
Correctness: compared with accuracy, 206-8
Correlation: in data processing, 115-7; in interpretation, 179-80; in fault interpretation, 185
Correlogram: correlated trace, 116
Cps: see Cycles per second
Crew: field, 4; land, 52-4; offshore, 69
Critical angle: in refraction, 21
Cross-correlation: comparison of two traces, 116
Curvature: of reflection, 30
Cuttings: see Drill
CVL: see Continuous velocity log
Cycles per second: seismic frequency, 106, also called cps, Hertz, H_z, 106

DAS: deconvolution after stack, 123
Data dump: printout, 129
Data exchange: for old data, 224-5

Data point: definition, 4
Data processing: general, 2, 4; usual, 93-129; special, 139-168; of old data, 229-230
Datum correction: for near-surface, 97-8
Datum plane: for near-surface correction, 97
DBS: deconvolution before stack, 123
Dead area: indicates salt dome, 236
Dead section: in marine cable, 76
Deconvolution: auto-correlation for, 117; in data processing, 122-4, 128, 129; signature deconvolution, 123-4
De-ghosting, 42
Demultiplex: rearrangement on tape, 93, 128; also called demux
Demux, 93
Density: effect on reflections, 6; in synthetic seismogram, 142
Depth: of reflection, 1; of offshore source and geophone, 132, 140; reliability of, 247
Depth controller: for marine cable, 76
Depth conversion: of seismic times, 139
Depth recorder: for water depth, 82
Depth section: section converted to depth, 136-7, 215
Detail: in program planning, 51
Detector: see Geophone
Deviation of line: effect on CDP, 16
Diazo print: section reproduction, 134
Diffraction: on compressed section, 138; hyperbolic pattern, 155-6; diffraction collapse migration, 157-9; in fault interpretation, 186-7; confused with anticline, 214
Diffraction collapse, 157-9
Digital: numerical data, 86-7
Digital processing: filtering, 107; use of signal theory, 115; tape, 229
Dip; causes of offset of reflection, 8; effect on CDP, 16; shown as color, 135; effect on refraction, 220; reliability of, 246-7
Dip Rejection: dip-selective processing, 126-7; also called moveout filter, 126
Display: of section, 134-5
Doghouse: recording room, 79; conditions in, 129, 130
Domain: time, frequency, 124-5
Doodlebug crew: seismic crew, 4
Doodlebugger: person on crew, 4
Doppler: in satellite navigation (radio), 73; doppler sonar (sound), 73
Doublet: reflection character, 179
Doubleton, reflection character, 179

Drape: over reef, 240, 242
Drill: description, 57-9; drill stem, 57; mast, pulldown, rotary table, slush pit, cuttings, 58; air drill, pumper drill, 58; puffer drill, 59
Duodecimal: number system, 88
Dynamite: as energy source, 56, 60

Echo-ranging: includes seismic, 5
Edit: of tapes, 128
Electrical field: effect on tape, 129
Energy: seismic waves, 4; frequency of, 106
Engineering Surveys: for well locations, 82-4
Eraser: type for sections, 177
Explosive: as seismic source, 78; explosive cord, 61
Exponential Notation: 89-90
Extended range shoran: see Shoran

Fan shooting: refraction, for salt domes, 222
Fault: in near-surface, 84; detected by velocity, 113; on compressed section, 138; general, 184-191; interpretation of, 217; in refraction, 222; need to define, 256
Field crew, 52-4
Field development: after discovery, 253-4
Field notes: of operations, 92
Field supervisor: 54
Field tape: in shooting, 4; in data processing, 128; retention of, 160
Filter: excluding some frequencies, 106-7; bandpass, 106; time variant, 108; in processing sequence, 128; on paper records, 227
Final tape: end of processing, 4; in processing sequence, 128; retention of, 160
First breaks: on record, 25; in near surface corrections, 97
Flat spot: gas indicator, 211-3, 217
Flattening: of horizon, 125-6
Floating point: gain recording, 90-1; for true amplitude, 151
Fold: degree of stack, 104
Formation: reflection tracking of, 181-4
Fourier: trace expressed as frequencies, 124-5
Frequency: of seismic range, 1; penetrations of, 107; in Fourier analysis, 124; shown as color, 135
Frequency display: records at different frequencies, 107, 129

Frequency domain: data expressed as frequencies, 124-5

Gain control: in recording, 90; in processing sequence, 128
Galvanometer: to make paper record, 227
Gamma ray log: on synthetic seismogram, 142
Gas gun: on land, 62; offshore, 78
Gather: single trace, 102; CDP, 102-3; CDP for velocity, 111
Geophone: detector, phone, pickup, seismometer, 4; description, 66-7; in velocity survey, 141; see also Hydrophone
Ghost: description, 41-3; de-ghosting, 42; detect by auto-correlogram, 117; diminished by deconvolution, 122
Glacial drift: resolved by flattening, 125; near-surface problem, 206
Gravity: exploration method, gravimeter, gravity meter, 85
Group: of geophones, 10-11; spacing of, 11; number of, 11; tapered, 11; string of geophones, 67
Gulf Coast, 238
Gulf of Mexico, 180-1
Gyrocompass: in satellite surveying, 73
Gyroscope: in inertial navigation, 73

Hertz, 106
High-pressure zone: detect by velocity, 113
Horizon: continuous reflection, 4, 175
Hot spot: see Bright spot
Hummingbird: field in Canada, 236
Hundred percent: complete subsurface coverage, 13; from selected shots, 100-102; single trace gather, 102
Hydrodynamic: tilts gas-liquid contact, 213
Hydrophone: marine geophone, 4, 75; also called pressure phone, pressure geophone, 75
H_z, 106

Identification: by velocity, 113; of reflection, 139, 175-6; by refraction velocity, 222
Indonesia, 239
Inertial navigation: in satellite surveying, 73
Inertial platform: in satellite surveying, 73
Integrated satellite system, 72
Interbed multiple, 40
Interface: convolved with signal, 119; in

synthetic sonic log, 147; reflection at, 181; many together, 182
Interference: in radio surveying, 74
Intermediate tape: in processing, 4; retention of, 160
Interpretation: sequence of, 215-7; of old and new data, 232
Interval map, 206, 217
Instrument cable: see Cable
Isochron: uses of word, 206
Isopach: 206

Jug: geophone, 4
Jug hustler: recording helper, 67

Lane: in radio surveying, 74
Large structures, 233-4
Library: of data, 224
Lignite: near-surface problem, 206
Line: of shooting, 4
Lithologic variation: effect on velocity, 48; on sections, 242-3
Log: of field operations, 79
Long range navigation: see Loran-C
Loop: in surveying, 55; to straighten cable, 77; in shooting and interpretation, 180-1; in interpretation, 216; in program planning, 250
Loran-C: satellite surveying, 73, 74; name from long range navigation, 73; RF (radio frequency), 74
Louisiana: 228
LVL: weathering, 29
LVL crew: for near-surface, 29, 65

Magnetic field: effect on tape, 129
Magnetic tape: see Tape
Magnetometer: exploration method, 85
Map: program, 3, 54; assignment, 53; on plane table, 54; "sloppy", 180; faults on, 191; general, 191-2; contouring on, 192-206; interval map, 206, 217; of old data available, 225
Marine cable: see Cable
Marine gas gun: see Gas gun
Maximum: gravity reading, 85
Maximum convexity: diffraction collapse migration, 159
Microfilm: of old records, 228
Midpoint: reflection from, 7
Migration: re-positioning reflections, 151-7, 256; wave equation, 159-60; comments on, 160-4; three-dimensional, 164-8, 215
Minimum: gravity reading, 85

Mis-tie: see Tie
Mix: on paper records, 227
Mobile station: in shoran surveying, 71
Modeling: synthetic sections, 148-51
Montana, 235-6, 246
Most tolerant position: for well location, 247-8
Moveout: as dip, 126; see also Normal moveout
Moveout filter, 126
Mud flat: shooting on, 81-2
Multiple: also called multiple reflection, 35; repeated reflection, 35-41; pegleg, 40; attenuation of, 105-6; detected by auto-correlation, 117; diminished by deconvolution, 122; removed by dip rejection, 127; on section, 133; on compressed section, 138; in synthetic seismogram, 145; confused with primary, 214
Multiple sources: like geophone group, 11
Multiplex: data arrangement on tape, 92; also mux, 93
Mute: remove parts of traces, 108-9, 128
Mux, 93

Navy Navigation Satellite System, 72-4
Near-surface corrections: of seismic times, 94-8; avoid by interval map, 206; provided by refraction, 218
Near traces, 228
Nine-track: tape format, 91
Nitrogen removal: for gas gun, 78
NMO: see Normal moveout
NNSS: satellite system, 72-4
Noise: reduced by geophone group, 10, 11; unwanted energy, 45-7; frequencies of, 106; most not diminished by deconvolution, 123
Noise arc: on migrated section, 159
Normalize: amplitude in stack, 104
Normal moveout: reflection curvature, 30-5, 103; effect of multiple on, 106; determines velocity, 109, 112; corrected on section, 132; in synthetic sonic log, 147; on paper records, 228; also called NMO, 35
North Dakota, 194, 201

Octal: number system, 88
Off center: of offshore shots, 231
Offset: of reflection point with dip, 8
Offshore crew, 69
Old data, 224-232

Operator: in signal theory, 115; in con-
volution, 122; in deconvolution, 122
Outlook: area in Montana, 235-6
Oxygen: for gas gun, 78

Participation: expense sharing seismic
program, 244
Pattern holes: 60-61
Peak: of trace, 23; in stack, 103-4; in
correlogram, 116; filled in, in VA, 134;
wiggle and VA, 135
Pegleg multiple, 40
Pencil: colored, for section marking,
176-7, 216
Penetration: of energy into earth, 109
Percent: degree of stack, 104
Permit man: on field crew, 54
Permitting: to shoot, 54
Phase: zero-phase wavelet, 48; in Fourier
analysis, 124
Phase-comparison: radio survey method,
74
Phone: see Geophone
Pick: in interpretation, 175, 216; on paper
records, 228
Pickup: see Geophone
Piezo-electric: hydrophone, 75
Pigtail: on geophone, 67
Pinchout: with synthetic seismogram,
146; not where it seems, 215; men-
tioned, 217, 242
Plane table: for surveying, 54
Plus or minus: significance of, 208
Polarity: of wavelet, 47; shown as color,
135; change at bright spot, 208
Pop: use of source, 4, 62, 132
Power spectrum: shows frequencies, 125
Pre-plots: for radio surveying, 71
Pressure geophone: see Hydrophone
Pressure phone: see Hydrophone
Primary: and multiple, 105
Processing sequence: 127-9
Program: of shooting, 2; map, 51
Programmed gain control: in recording,
90
Program planning: for shooting, 2, 51,
250-4; detail, 51; place of old data in,
226, 231; for large structures, 233-4; for
small features 234-5; for continuing
program, 252-3; in discovered field,
253-4
Propane: in gas gun, 78
Proprietary: program for sale by contrac-
tor, 244
Puffer drill, 59

Pulldown: see Drill
Pull-down: by velocity, under reef, 241
Pull-up: by velocity, under reef, 240
Pumper drill, 58

Radar: compared with seismic, 5; signal
theory developed for, 115
Radio frequency: in loran-C, 74
Radio transmitter: in telemetry, 81; in
sonobuoy, 222-3
Rainbow-Zama: area in Alberta, 179
Ray path: of seismic energy, 6; in wave
equation migration, 160; ray tracing in
modeling, 149
Ray tracing: see Ray path
Receiver: geophone, 5
Record: seismic record of one shot, 22-26,
103; old, paper, 226-231
Recording: on land, 67-8; offshore, 79-80
Recording format: 91-2
Record quality: 256
Record section: also called section, seis-
mic section, 4, 26; description, 26-8,
132-3; from final tape, 128; forms of,
134-8; of synthetic sonic logs, 147-8;
from old records, 229-230
Reef: detected by flattening, 125; detec-
tion of, 217, 239-41; as stratigraphic
trap, 242
Reference marks: on sonic log, 142
Reflection: compared with refraction, 1; of
sound, 4, 5-9, 25; of light, 218
Reflection coefficient: in synthetic seis-
mogram, 143-5
Reflection time, 4
Reflectivity: at interface, to make synthetic seismogram, 122
Refraction: compared with reflection, 1; of
sound, 19-22, 218-22; for deep horizons,
22; for near-surface, 22, 97; detected by
sonobuoy, 222-3; for salt domes, 236
Regional control, 250
Regional information: from old data, 225
Reliability: of seismic data, 246-8
Residual statics, 128
Reverberation: in marine shooting, 39;
singing, or ringing record, 39; di-
minished by deconvolution, 122
Roll-along: see Common depth point
Root mean square: also RMS, 113; seismic-
ally detected velocity, 113, 128; for
depth conversion, 140
RF: in loran-C, 74
RMS: see Root mean square

Saddle: reflection character, 179

Sale: of data, 244
Salt domes: exploration for, 180-1, 222, 236-9; fault pattern, 190
Salt solution, 235-6
Saskatchewan, 236
Satellite: survey, 72-4
Scale: of section, 132, 135-7
Scurry Field, Texas, 233
Sea bottom: on section, 133
Section: see Record section
Sedimentary basin, 6
Seismic definition, 1; survey, 2; energy, 4; line, 4; waves, 4; see Crew, Record section
Seismogram: word, 142-3
Seismograph crew, see Crew
Seismometer: see Geophone
Seven track: tape format, 92
Shadow zone: around bright spot, 211
Shallow water: shooting in, 81-2
Sheepherder anticline, 233
Shooting: seismic field operation, 4
Shoran: name from short range navigation, 71; XR shoran, 71; extended range shoran, 71; radio surveying, 71-2
Short range navigation: see Shoran
Shot hole: for explosive source, 4, 56-61
Shot point: location of source, 4
Shot point seismometer: near hole geophone, 25, 94; also called SPS, up-hole jug, 25, 94
Side-scan sonar: sea bed record, 82-4
Signal theory: source of terms, 4; in data processing, 115
Signal: seismic energy put into earth, 80-1
Signal to noise ratio: wanted to unwanted energy, 46; affected by mute, 109
Signature: of marine source, 80-1
Signature deconvolution: see Deconvolution
Singing: see Reverberation
Single-ended spread: see Spread
Single-ender: see Spread
Single fold: complete subsurface coverage, 12
Single trace gather: see Gather
"Sloppy" map in interpretation, 180-1
Small features: see Program planning
Smear: of CDP with dip, 17
Sonic log: to make synthetic seismogram, 122, 142; in velocity survey, 141-2
Sonobuoy: see Refraction
Sound: in seismic exploration, 5
Source: dynamite, 56, 60; pattern holes, 60-1; surface explosives, 61; explosive

cord, 61; gas gun, land, 62; weight drop, 62-3; vibrator, 63-5; tuned air gun array, 77; air gun, 77-8; gas gun, offshore, 78; steam, 78; explosives offshore, 78; electrical spark, 84
Spike: at velocity interface, 119-20; goal of deconvolution, 122
Split spread: see Spread
Spread: of instrument cable, 11-4; single-ended, single-ender, 12; split, 13
SPS: see Shot point seismometer
Stack: percents, 14; combining of traces, 102, 103-4, 128, 132; multiple attenuation, 106; reduced by mute, 109; in velocity determination, 112; stacked tape, 128
Static corrections: for near-surface, 98-100; also statics, 99; residual statics, 128
Steam: as seismic source, 78
Stratigraphic traps: exploration for, 241-2
Stratigraphy: shown on sections, 213-4
Streamer, see Cable
String: see Group
Structure: in near-surface, 84; on compressed section, 138
Subsurface coverage, 12
Surf: shooting in, 81-2
Surface explosives: seismic source, 61
Surveying: on land, 2, 54-6; offshore, 70-4; pre-plots, 71; shoran, 71-2; satellite, 72-4; loran-C, 74; problems, 244-6
Sweep: of vibrator, 63
Syncline: migration of, 154-5
Synthetic seismogram: made from well logs, 122, 142-6; in reflection identification, 176; in interpretation, 216; to detect lithologic change, 243
Synthetic sonic log: from seismic data, 146-8

Tail buoy: on marine cable, 76; appearance, 77
Takeout: connector on cable; 65
Tape: also magnetic tape, 4; seven track, nine track, twenty-one track formats, 91-2; distortion of, 129; storage of, 129-31
Tape recorder: in field recording, 67
Tapered group, see Group
Tape storage, 129-31
Telemetry: for geophones, 81; sonobuoy, 222-3
Texas, 233
Theodolite: survey instrument, 54, 82

Three dimensional migration, 164-8, 215
3-D shooting: 168-73, 256; for fault interpretation, 191
Three-way fix: in shoran surveying, 72
Tie: in surveying, 55, 246; in migration, 162-4; in interpretation, 180-1, 216; in fault interpretation, 185
Time: of reflection, 175
Time break: instant of shot, 25
Time, conversion to depth: with NMO velocity, 112, 137; depth section, 136-7, 215; with assumed velocities, 139-40; in horizon identification, 176; mentioned, 2
Time domain: data expressed as times, 124-5
Time interval map: use of, 206, 217
Time variant filter, 108, 128
Timing line: to indicate times on section, 25
Topography: on section, 133; indicates salt dome, 236
Trace: shows geophone motion, 4, 23; produced by interfaces, 19; as combination of cosine waves, 124; in section, 132; character of, 179
Trade service: for data exchange, 224
Transcribe: paper to tape, 225-6, 229
Transit: survey instrument, 54, 82
Transit: name of satellite system, 72
Trough: of trace, 23; in stack, 103-4; in correlation, 116; on section, 135
True amplitude: relative strengths on section, 128, 151, 211
t_{uh}: see Uphole
Tuned air gun array: see Air gun
TVF: see Time variant filter
Twenty-one track: tape format, 92

Unconformity: in rock layers, 242
Update: in satellite, 73-4
Uphole: break, 25; jug, 25, 94; time, 94; t_{uh}, 94

VA: variable area section, 135
Variable area: section display, 134
Variable density: section display, 134
VD: variable density section, 135
Velocimeter: in satellite survey, 73
Velocity: of sound through rock, 5; effect on reflection, 6; range of, 6; in normal moveout, 35; variations, 48-50; of multiple, 106; determination of, 109-114,

139-140; interface, 119; shown as color, 135; for depth section, 137; for synthetic seismogram, 142; in reflection identification, 176; effect of fluid in rock on, 209; increase with compaction, 218; determined by refraction, 221-2; of water, 231; in salt domes, 238; of reef, 239-241; in lithologic change, 243; effect on reliability, 246-7
Velocity anomaly: confused with structure, 215; pull-up, 240; pull-down, 241
Velocity geophone: land geophone, 67
Velocity survey: for synthetic seismogram, 122; for depth section, 137; in well, 140-2; for reflection identification, 176
Velocity survey crew, 141
Vibrator: seismic source, 63-5
Vertical exaggeration: of section, 135-7; in compressed section, 137-8; confusing, 215
Vessel: seismic boat, 69-70

Wabamun: formation in Alberta, 179
Water depth recorder: compared with seismic, 5
Wave equation migration: see Migration
Wavelet: unit of reflection, 47-8; zero-phase, 48; shown by auto-correlation, 117; convolved with interfaces, 120-2; in deconvolution, 122; combined in reflection, 183; delay in, 207; in lithologic change, 242
Weathering: near-surface layer, 29-30; shoot below, 94; measured by refraction, 220; also called W_x, 30
Weight drop: seismic source, 62-3
Well location: by satellite survey, 74; based on faults, 190; at top of feature, 254-5
Well log: compared with synthetic sonic log, 147
Well ties: in program planning, 250, 252-3
Wiggle trace: section display, 134
Williston Basin, 194, 201, 235-6, 250
Wobble: uncertainty of seismic data, 192
W_x: see Weathering
Wyoming, 233

XR shoran: see Shoran
X-Y coordinates: in radio surveying, 72

Zero-phase wavelet: see Wavelet

Date Due

M